Geometry
A Self-Teaching Guide

Geometry
A Self-Teaching Guide

Steve Slavin
Ginny Crisonino

WILEY

John Wiley & Sons, Inc.

ISBN 0-471-38634-0

Printed in the United States of America

10   9   8   7   6   5   4   3   2   1

Contents

Introduction

The best thing about this book is that you don't need to know a whole lot of math to be able to read it. If you can add, subtract, multiply, and divide, then you're all set. Before you know it, you'll know all about every geometric shape from circles and squares to cubes, spheroids, and cones.

Like readers of all the other self-teaching guides in this series, you'll work your way through this book problem by problem. We'll be right there with you every step of the way. And you'll find a full solution to each problem in the book. So you can check your work, go back over things that you need to review, and work at your own pace.

The chances are, much of what you'll see in the first three chapters will be quite familiar. At the beginning of each chapter is a pretest, so if you score well, you may skip part or all of the chapter. Even so, for the sake of review, you may want to work your way through these chapters.

You'll notice that there are several self-tests within each chapter, so you can constantly monitor your progress. It will be up to you to be completely honest with yourself. You'll need to keep asking: Self, do I really understand everything in this section? If the answer is no, you'll need to reread that section and redo each problem. Generally, you should do this if you get more than one wrong answer on a self-test.

Some people love geometry, while others hate it. We hope you'll have as much fun reading our book as we did writing it. But then again, we already knew this stuff before we started. By the time you've finished reading this book, maybe you'll be ready to write your own book.

Ginny Crisonino has taught mathematics at Union County College in Cranford, New Jersey, since 1983. Together with her fabulous coauthor, Steve Slavin, she has written *Precalculus: A Self-Teaching Guide* (published by Wiley) and *Basic Mathematics* (πr^2 Publishing Company), a college text now in its second edition. Steve Slavin taught economics for 30 years and has authored or coauthored 14 books in mathematics and economics, including *Economics* (McGraw-Hill), a college text now in its seventh edition, and *All the Math You'll Ever Need* (Wiley), now in its second edition and a Literary Guild selection.

1 The Basics

Chapter 1 reviews some of the basic concepts of geometry. When you've completed this chapter, you should be able to work with the following concepts:

- points and lines
- angles
- polygons

At the beginning of each chapter we'll give you a pretest. If you get a perfect score on a pretest without peeking at the answers that follow, it's probably all right to skip that chapter. Before we begin chapter 1, try the following pretest to get an idea of how much review you need, if any. If you get a perfect score on the pretest, you may skip chapter 1 and go directly to chapter 2.

PRETEST

1. Find the distance between the following sets of points.

   a. (1,9) and (3,7)   b. (−2,6) and (6,−2)

2. Find the midpoint of both line segments in the previous problem.

3. List all the line segments for the following line.

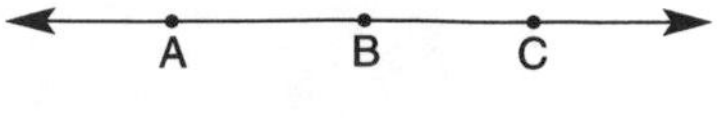

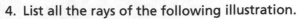

4. List all the rays of the following illustration.

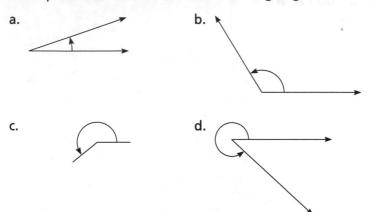

5. Use a protractor to measure the following angles.

a.

b.

c.

d.

6. Without using a protractor, state the values of the angles in the following figure.

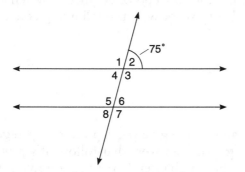

7. Find the value of x in the following figure.

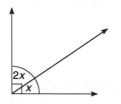

8. State whether the following angles are acute, obtuse, right, or straight.

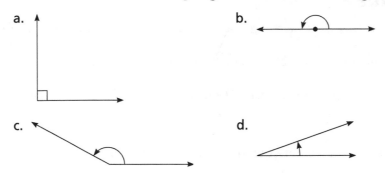

a.

b.

c.

d.

9. What's the sum of the measure of the angles of a hexagon?

10. What kind of quadrilateral has four right angles but not all sides are equal?

ANSWERS

1. a. $d = \sqrt{(3-1)^2 + (7-9)^2} = \sqrt{4+4} = \sqrt{8} = 2\sqrt{2} \approx 2.83$

   b. $d = \sqrt{(6+2)^2 + (-2-6)^2} = \sqrt{64+64} = \sqrt{128} = 8\sqrt{2} \approx 11.31$

2. a. *Midpoint:*
   $$\left(\frac{1+3}{2}, \frac{9+7}{2}\right) = (2,8)$$
   b. *Midpoint:*
   $$\left(\frac{-2+6}{2}, \frac{6+-2}{2}\right) = (2,2)$$

3. $\overline{AB}$ $\overline{AC}$ $\overline{BC}$ 4. $\overleftrightarrow{ZX}$ $\overleftrightarrow{ZY}$

5. a. 20°   b. 120°   c. 220°   d. 320°

6. $\angle 1 = \angle 3 = \angle 5 = \angle 7 = 105°$ $\angle 2 = \angle 4 = \angle 6 = \angle 8 = 75°$

7. $x + 2x = 90°$

   $3x = 90°$

   $x = 30°$

8. a. right   b. straight   c. obtuse   d. acute

9. $(6-2)180° = 4(180°) = 720°$ 10. rectangle

Points and Lines

Let's start with something you probably know. Fill in the rest of this sentence:

The shortest distance between two points is _____.

Did you know the answer? *The shortest distance between two points is a straight line.*

Example 1:
Find the distance between the points (0,4) and (4,2).

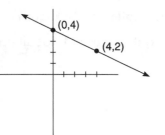

Solution:
The distance formula is used to find the distance between any two points, (x_1, y_1) and (x_2, y_2).

Distance formula:

$$d = \sqrt{(x_2 - x_1)^2 + (y_2 - y_1)^2}$$

The distance from (0,4) to (4,2) is the same distance as from (4,2) to (0,4), so it doesn't matter which point is designated as the first point (x_1, y_1) or the second point (x_2, y_2). We'll prove this by calculating the distance both ways.

Suppose we let (0,4) be the first point and (4,2) be the second point; then $x_1 = 0$, $y_1 = 4$, and $x_2 = 4$, $y_2 = 2$. If we substitute these values into the distance formula and simplify, it will look like this: $d = \sqrt{(x_2 - x_1)^2 + (y_2 - y_1)^2} = \sqrt{(4 - 0)^2 + (2 - 4)^2} = \sqrt{(4)^2 + (-2)^2} = \sqrt{16 + 4} = \sqrt{20} = \sqrt{4(5)} = 2\sqrt{5} \approx 4.47$. The distance between the given points is approximately 4.47 units.

Now we'll reverse the points and call (4,2) the first point and (0,4) the second point. Then $x_1 = 4$, $y_1 = 2$, and $x_2 = 0$, $y_2 = 4$. If we substitute these values into the distance formula and simplify, it will look like this: $d = \sqrt{(x_2 - x_1)^2 + (y_2 - y_1)^2} = \sqrt{(0 - 4)^2 + (4 - 2)^2} = \sqrt{(4)^2 + (2)^2} = \sqrt{16 + 4} = \sqrt{20} = 2\sqrt{5} \approx 4.47$. No matter which one we call the first point and which one we call the second point, the distance is still the same.

Example 2:

Find the distance between the points (–1,2) and (2,–1).

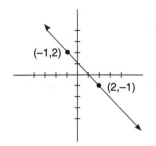

Solution:

We can choose either point to be the first point; the other point is the second point. We'll call (–1,2) the first point and (2,–1) the second point. Therefore, $x_1 = -1$, $y_1 = 2$ and $x_2 = 2$, $y_2 = -1$.

$d = \sqrt{(2 - -1)^2 + (-1 - 2)^2}$	Substitute the values into the distance formula.
$d = \sqrt{(2 + 1)^2 + (-3)^2}$	Minus a negative becomes plus a positive.
$d = \sqrt{(3)^2 + (-3)^2}$	
$d = \sqrt{9 + 9}$	
$d = \sqrt{18} = \sqrt{9(2)} = 3\sqrt{2} \approx 4.24$	

Suppose we wanted to find the coordinates of the halfway mark or midpoint of the line that connects the points in example 1. Would you know to use the midpoint formula? Given two points (x_1, y_1) and (x_2, y_2), use the following formula to find the midpoint. As in the distance formula, it doesn't matter which point you call the first point and which point you call the second point; the midpoint will still be the same.

Midpoint formula:

$$\left(\frac{x_1 + x_2}{2}, \frac{y_1 + y_2}{2} \right)$$

Example 3:

Find the midpoint of the line segment that connects the points (0,4) and (4,2) in example 1.

Solution:

It doesn't matter which point we call the first point (x_1, y_1). Let's call $(0,4)$ the first point, which makes $(4,2)$ the second point. Therefore, $x_1 = 0$, $y_1 = 4$ and $x_2 = 4$, $y_2 = 2$. Substitute these values into the midpoint formula and simplify.

Midpoint:

$$\left(\frac{x_1 + x_2}{2}, \frac{y_1 + y_2}{2}\right) = \left(\frac{0 + 4}{2}, \frac{4 + 2}{2}\right) = \left(\frac{4}{2}, \frac{6}{2}\right) = (2,3)$$

Example 4:

Find the midpoint of the line segment that connects the points $(-1,2)$ and $(2,-1)$ in example 2.

Solution:

Let's choose $(-1,2)$ to be the first point; then $(2,-1)$ has to be the second point. Therefore, $x_1 = -1$, $y_1 = 2$ and $x_2 = 2$, $y_2 = -1$. Substitute these values into the midpoint formula and simplify.

Midpoint:

$$\left(\frac{x_1 + x_2}{2}, \frac{y_1 + y_2}{2}\right) = \left(\frac{-1 + 2}{2}, \frac{2 + -1}{2}\right) = \left(\frac{1}{2}, \frac{1}{2}\right)$$

Line Segment

Here's our next definition: *A line segment is a section of a line between two points.*

Using that definition very carefully, see if you can identify all three line segments in the following line.

The three line segments in this line are $\overline{DE}$, $\overline{EF}$, and $\overline{DF}$. A line drawn over two capital letters will stand for line segment. That's right, $\overline{DF}$ is a line segment. Remember our definition: *A line segment is a section of a line between two points*— even if those points happen to be located at the two extremes of the line, with other points in between, and even if the section of the line is the entire line. The order in which we list the letters doesn't matter. $\overline{DE}$ is considered to be the same line segment as $\overline{ED}$.

Example 5:

Identify all the possible line segments of the following line.

Solution:

$\overline{AB}, \overline{AC}, \overline{AD}, \overline{AE}, \overline{BC}, \overline{BD}, \overline{BE}, \overline{CD}, \overline{CE}, \overline{DE}$

Ray

Finally we have a ray. *A ray has an end point at one end of a line and extends indefinitely in the other direction.* An arrow indicates the direction. The following is a ray.

This ray can be written in the following ways. We'll put an arrow over a ray, with the arrow pointing in the same direction as the original ray.

$\overrightarrow{NM}, \overrightarrow{NO}$

Example 6:

List all the possible rays from the illustration below.

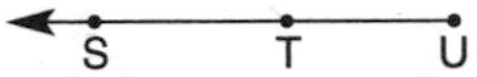

Solution:

$\overleftarrow{UT}, \overleftarrow{US}$

SELF-TEST 1

1. Find the distance of the straight line that connects these points:

    a. (3,9) and (7,13)   b. (−5,3) and (3,7)

2. Find the coordinates of the midpoint of the straight lines between these points:

    a. (3,9) and (7,13)   b. (−5,3) and (3,7)

3. Identify all of the line segments of each of the following lines.

    a. 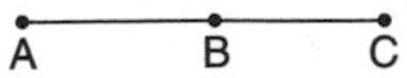 A   B   C   b. X   Y   Z

    c. A   B 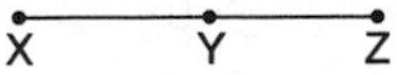   C    D    d. J   K   L   M

4. List all the possible rays of the following lines.

    a. 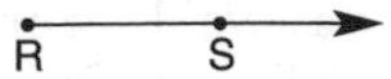 R   S   b. A   B

    c. 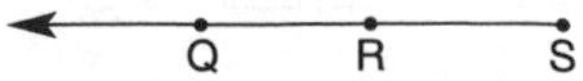 Q   R   S   d. 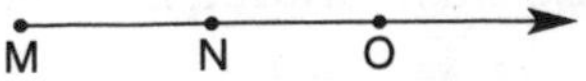 M   N   O

1. $d = \sqrt{(x_2 - x_1)^2 + (y_2 - y_1)^2}$

   a. $x_1 = 3$, $y_1 = 9$, and $x_2 = 7$, $y_2 = 13$

   $d = \sqrt{(7-3)^2 + (13-9)^2} = \sqrt{(4)^2 + (4)^2} = \sqrt{16 + 16} = \sqrt{32} = \sqrt{16(2)} = 4\sqrt{2} \approx 5.66$

   b. $x_1 = -5$, $y_1 = 3$, and $x_2 = 3$, $y_2 = 7$

   $d = \sqrt{(3 - -5)^2 + (7-3)^2}$

   $= \sqrt{(3+5)^2 + (7-3)^2} = \sqrt{(8)^2 + (4)^2} = \sqrt{64 + 16} = \sqrt{80} = \sqrt{16(5)} = 4\sqrt{5} \approx 8.94$

2. *Midpoint:*

   $\left(\dfrac{x_1 + x_2}{2}, \dfrac{y_1 + y_2}{2} \right)$

   a. $x_1 = 3$, $y_1 = 9$, and $x_2 = 7$, $y_2 = 13$
   Midpoint:

   $\left(\dfrac{3+7}{2}, \dfrac{9+13}{2} \right) = \left(\dfrac{10}{2}, \dfrac{22}{2} \right) = (5,11)$

   b. $x_1 = -5$, $y_1 = 3$, and $x_2 = 3$, $y_2 = 7$
   Midpoint:

   $\left(\dfrac{-5+3}{2}, \dfrac{3+7}{2} \right) = \left(\dfrac{-2}{2}, \dfrac{10}{2} \right) = (-1,5)$

3. a. $\overline{AB}$ $\overline{AC}$ $\overline{BC}$ b. $\overline{XY}$ $\overline{XZ}$ $\overline{YZ}$

   c. $\overline{AB}$ $\overline{AC}$ $\overline{AD}$ $\overline{BC}$ $\overline{BD}$ $\overline{CD}$ d. $\overline{JK}$ $\overline{JL}$ $\overline{JM}$ $\overline{KL}$ $\overline{KM}$ $\overline{LM}$

4. a. $\overrightarrow{RS}$ b. $\overrightarrow{BA}$ c. $\overleftarrow{SQ}$ $\overrightarrow{RQ}$ d. $\overrightarrow{MN}$ $\overrightarrow{MO}$

Angles

If two straight lines meet or cross each other at a point, an angle is formed. The point where the lines meet is called the vertex of the angle; the sides are called the rays of the angle. For example, in the following angle, the vertex is at the point B, and the rays are $\overrightarrow{BA}$ *and* $\overrightarrow{BC}$.

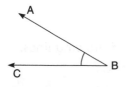

We use $\angle$ as the symbol for an angle. The angle shown above can be called $\angle ABC$ (where the middle letter names the vertex) or simply $\angle B$.

Angles are usually measured in degrees, but as we'll see in chapter 3, they can also be measured in radians. Perhaps the most familiar angle is the right angle. The symbol for the right angle is a square drawn at the vertex. A right angle is formed when two perpendicular lines intersect, forming a 90° angle. The following is a drawing of a right angle.

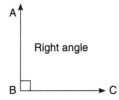

Most people can easily recognize right angles, but what about the following angle?

Would you be able to look at it and tell us it's 67°? No one is that good, not even the authors. We can identify some angles just by looking at them. So the obvious question is: How do we find the measure of an angle? The answer is: Use a protractor.

Let's start by looking at a circle. A circle has 360°.

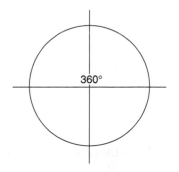

If we divide the circle into four equal parts formed by the intersection of two perpendicular lines, each part is 90°, a right angle.

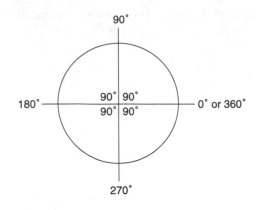

But how do we measure angles that aren't circles or right angles? The protractor measures the number of degrees of an angle. Any angle that moves counterclockwise is positive; any angle that moves clockwise is negative. For the remainder of this section, we'll work only with positive angles. Go back three illustrations to the 67° angle. Looking at the angle, you can see it's between 0° and 90°. If you place your protractor over the angle with the vertex lined up with the line on your protractor, you'll find that the angle is 67°. *Angles between 0° and 90° are called acute angles.* The following figure shows an acute angle.

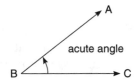

Angles larger than 90° but smaller than 180° are called obtuse angles. The following figure shows an obtuse angle.

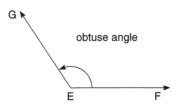

Another type of angle is the straight angle, which measures 180°. As you can see, a straight angle is a straight line.

We measure angles with the use of a protractor. As we mentioned, certain angles are so common that we know what they are just by looking at them. Angles of 90° and 180° are easily recognized. In addition, angles of 30°, 45°, 60°, and 270° are common.

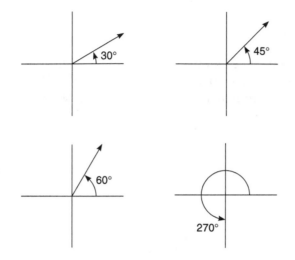

Example 7:

Use a protractor to measure the following angles and label them as acute, right, obtuse, or straight.

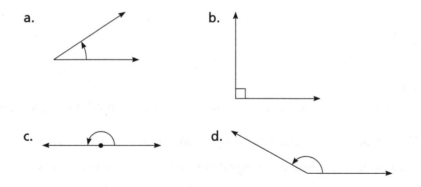

Solution:

    a. This angle is approximately 32°, which makes it an acute angle.

    b. This angle is 90°, which makes it a right angle.

    c. This angle is 180°, which makes it a straight angle.

    d. This angle is approximately 150°, which makes it an obtuse angle.

Angles f and g, shown in the following illustration, are *adjacent angles.* You'll notice that they have the same vertex. Therefore: *if two angles have the same vertex and are adjacent to each other, they are adjacent angles.*

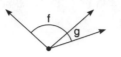

The sum of the two adjacent angles shown in the following illustration is 90°. *Any two angles whose sum is 90° are complementary angles.* So angles DEF and FEG are complementary. However, complementary angles do not have to be adjacent.

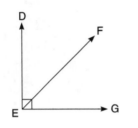

Example 8:

Find the value of *x* in the following illustration.

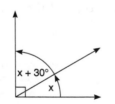

Solution:

These angles are complementary, so we know the sum of the angles is 90°.

$x + 30° + x = 90°$	Write this information in an equation form.
$2x + 30° = 90°$	Solve the equation by combining like terms.
$2x = 60°$	Subtract 30° from both sides of the equation.
$x = 30°$	Divide both sides of the equation by 2.

The sum of the following two adjacent angles is 180°. *Any two angles whose sum is 180° are supplementary angles.* Angles r and s are supplementary.

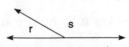

In the following figure, are angles w and v supplementary? They certainly are. But please note that like complementary angles, supplementary angles do not have to be adjacent. All that matters is the sum. Hold that thought for a few minutes and we'll show you some supplementary angles that are not adjacent.

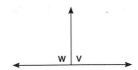

When two lines intersect each other, four angles are formed, and their sum is 360°.

$a° + b° + c° + d° = 360°$

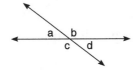

The sum of the adjacent angles is 180°.

$a° + b° = 180°$

$c° + d° = 180°$

$a° + c° = 180°$

$b° + d° = 180°$

Angles that have the same degrees are called equivalent angles. The symbol for equivalent angles is ≅.
Angles diagonally opposite each other are called vertical angles, and they are equivalent.

$\angle a \cong \angle d$

$\angle b \cong \angle c$

Two lines perpendicular to each other make up a special case. These lines form four right angles, as shown in the next figure.

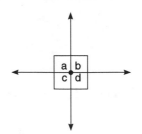

$\angle a \cong \angle b \cong \angle c \cong \angle d \cong 90°$

$\angle a + \angle b + \angle c + \angle d = 360°$

The following illustration is a transversal. *A transversal is a line that intersects two other lines.* The transversal intersects the two lines, forming eight angles numbered 1 through 8 in the figure.

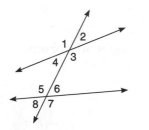

Here's another transversal. This time it crosses parallel lines.

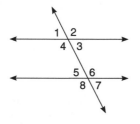

This transversal creates several different kinds of angles: corresponding, interior, and exterior. We will use the figure above to illustrate these three kinds of angles.

Corresponding angles are two angles in corresponding positions relative to two lines. The following pairs of angles are corresponding angles.

$\angle 1$ and $\angle 5$ $\angle 2$ and $\angle 6$ $\angle 4$ and $\angle 8$ $\angle 3$ and $\angle 7$

Corresponding angles are congruent (have equal measure).

Interior angles are angles between a pair of lines crossed by a transversal. The interior angles are $\angle 3$, $\angle 4$, $\angle 5$, and $\angle 6$.

Alternate interior angles are two nonadjacent interior angles on opposite sides of a transversal. The following pairs are alternate interior angles.

$\angle 4$ and $\angle 6$ $\angle 3$ and $\angle 5$

Alternate interior angles are congruent.

Exterior angles are angles outside a pair of lines crossed by a transversal. Exterior angles are $\angle 1$, $\angle 2$, $\angle 7$, and $\angle 8$.

The following are pairs of alternate exterior angles.

∠1 and ∠7   ∠2 and ∠8

Alternate exterior angles are congruent.
 The following pairs of angles are supplementary (they add up to 180°).

∠1 and ∠2   ∠1 and ∠4   ∠2 and ∠3   ∠3 and ∠4

∠5 and ∠6   ∠5 and ∠8   ∠6 and ∠7   ∠7 and ∠8

If lines D and E are parallel, as shown in the following illustration, then ∠d = ∠g = ∠h = ∠k and ∠e = ∠f = ∠i = ∠j.

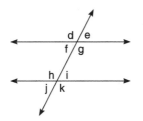

Example 9:

Find the values of the angles in the following illustration if ∠1 = 135°. Assume the lines, f and g, crossed by the transversal are parallel.

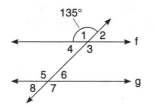

Solution:
 ∠1 and ∠3 are vertical angles, which are congruent; therefore, ∠3 = 135°.
 ∠1 and ∠2 are supplementary angles; therefore, ∠2 = 180° − 135° = 45°.
 ∠1 and ∠4 are also supplementary; therefore, ∠4 = 45°.
 ∠4 and ∠6 are alternate interior angles, which are congruent; therefore, ∠6 = 45°.
 ∠5 and ∠6 are supplementary; therefore, ∠5 = 135°.
 ∠6 and ∠8 are vertical angles; therefore, ∠8 = 45°.
 ∠5 and ∠7 are vertical angles; therefore, ∠7 = 35°.
 ∠1 ≅ ∠3 ≅ ∠5 ≅ ∠7 = 135°.
 ∠2 ≅ ∠4 ≅ ∠6 ≅ ∠8 = 45°.

SELF-TEST 2

1. Identify the vertex of each of the following angles.

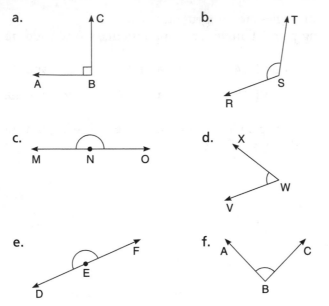

2. State whether each of the angles in the previous question is acute, obtuse, right, or straight.

3. Each of the following pairs of angles can be described as adjacent, complementary, or supplementary. Use one or two of these descriptions for each angle.

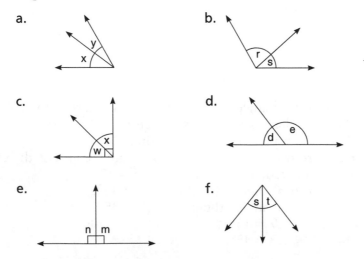

4. Use the following figure to answer these questions.

   a. What is the sum (in degrees) of angles a, b, c, and d.

   b. What is the sum of angles a and b?

   c. What is the sum of angles c and d?

   d. What is the sum of angles a and d?

   e. What is the sum of angles b and c?

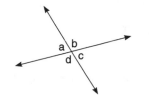

5. Find the degree of angles k, l, m, n, o, p, and q. Assume the lines cut by the transversal are parallel.

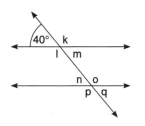

6. Find the degree of angles a, c, d, e, f, g, and h. Assume the lines cut by the transversal are parallel.

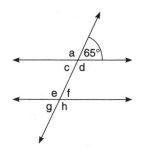

7. Find the value of x in the following illustration.

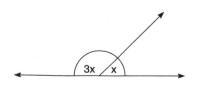

1. a. B   b. S   c. N   d. W   e. E   f. B

2. a. right   b. obtuse   c. straight

    d. acute   e. straight   f. acute

3. a. adjacent   b. adjacent

    c. adjacent, complementary   d. adjacent, supplementary

    e. adjacent, supplementary   f. adjacent

4. a. $360°$ b. $180°$ c. $180°$ d. $180°$

5. $k = 140°$ $1 = 140°$ $m = 40°$ $n = 40°$ $o = 140°$ $p = 140°$ $q = 40°$

6. $a = 115°$ $c = 65°$ $d = 115°$ $e = 115°$ $f = 65°$ $g = 65°$ $h = 115°$

7. $3x + x = 180°$

   $4x = 180°$

   $x = 45°$

Polygons

A polygon is a closed planar figure that is formed by three or more line segments that all meet at their end points; there are no end points that are not met by another end point.

Two of the most common polygons are triangles (to which we've devoted all of the next chapter) and quadrilaterals.

Quadrilaterals are four-sided figures. Shown below are the four most common quadrilaterals.

A square is a quadrilateral that consists of all right angles and equal sides.

A rectangle is a quadrilateral that consists of four right angles where opposite sides are parallel and equal in length.

A trapezoid is a quadrilateral with exactly one pair of parallel sides.

A parallelogram is a quadrilateral with both pairs of opposite sides parallel.

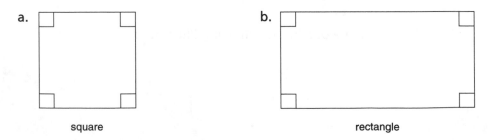

a. square   b. rectangle

c. 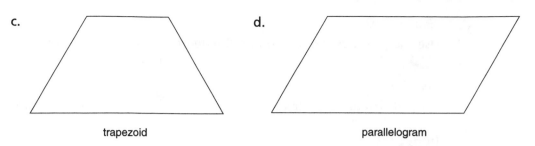 d.

trapezoid   parallelogram

All quadrilaterals contain 360°. The square and the rectangle each contain four angles of 90°.

The trapezoid and the parallelogram each contain two angles of more than 90°. A few pages back we said that supplementary angles could be adjacent or nonadjacent. Trapezoids contain two sets of nonadjacent supplementary angles. So, too, do parallelograms. The following illustrations show an example of each.

How many degrees are angles A, B, and C in each of the following polygons?

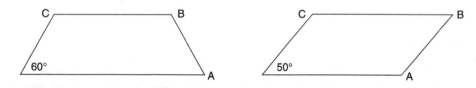

In the figure, the trapezoid contains angles A (60°), B (120°), and C (120°). The parallelogram contains angles A (130°), B (50°), and C (130°).

So far we've covered just quadrilaterals. Can you guess what a five-sided polygon is called? It's called a pentagon. Of course, the most famous pentagon is the head-quarters of the U.S. Department of Defense, a five-sided building just outside Washington, D.C. Try to guess what a six-sided polygon is called. It's a hexagon. We'll tell you the next four—a seven-sided polygon is a heptagon; one with eight sides is an octagon; one with nine sides is a nonagon; and one with ten sides is a decagon.

We've already seen that the sum of the four angles of a quadrilateral is 360°. Now we'll give you the formula that provides the sum of all the angle measurements in any polygon:

The sum of all angles in an n-gon = $(n-2)180°$, where n is the number of angles (or sides) in the polygon.

Example 10:

Use the previous formula to find the degrees of a quadrilateral.

Solution:

A quadrilateral has four sides, so $n = 4$. Substitute 4 for n in the following formula.

$(n - 2)180°$

The sum of all the angles in a quadrilateral $= (4 - 2)180° = (2)180° = 360°$.

Example 11:

Find the sum of the angles in a decagon.

Solution:

A decagon has ten sides, so $n = 10$. Substitute 10 for n in the following formula.

$(n - 2)180°$

The sum of all the angles in a decagon $= (10 - 2)180° = (8)180° = 1,440°$.

SELF-TEST 3

1. How many angles are contained in a square, and how many degrees is each angle?

2. How many angles are contained in a rectangle, and how many degrees is each angle?

3. How many pairs of supplementary angles are contained in a square and in a rectangle?

4. a. How many pairs of supplementary angles are contained in a trapezoid?

   b. Are they adjacent or nonadjacent?

5. a. How many pairs of supplementary angles are contained in a parallelo-gram?

   b. Are they adjacent or nonadjacent?

6. Find the angles A, B, and C of these four polygons.

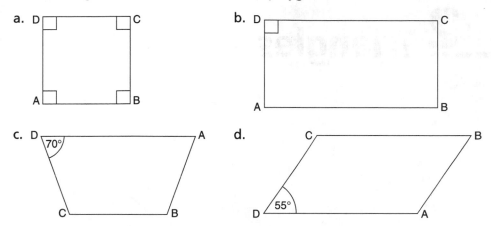

7. Find angles D, E, F of these two polygons.

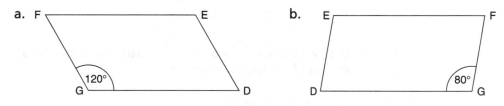

8. Find the sum of the angles in a hexagon (six-sided polygon).

9. Find the sum of the angles in an octagon (eight-sided polygon).

ANSWERS

1. 4 angles; 90°   2. 4 angles; 90°   3. 2; 2

4. a. 2   b. nonadjacent   5. a. 2   b. nonadjacent

6. a. A = 90°   B = 90°   C = 90°   b. A = 90°   B = 90°   C = 90°

   c. A = 70°   B = 110°   C = 110° d. A = 125°   B = 55°   C = 125°

7. a. D = 60°   E = 120°   F = 60°   b. D = 80°   E = 100°   F = 125°

8. $(n - 2)180° = (6 - 2)180° = (4)180° = 720°$

9. $(n - 2)180° = (8 - 2)180° = (6)180° = 1,080°$

2 Triangles

In this chapter we'll take a closer look at triangles. When you've completed this chapter, you should be able to work with:

- the Pythagorean theorem
- similar and congruent triangles
- perimeter and area of triangles
- applications of triangles

PRETEST

Let's find out how much you know about triangles before we move into the chapter. If you really know a lot, you'll be able to skip one or two sections. And if you know even more, then we'll let you skip most of the chapter.

1. How would you describe each of these triangles? Choose from: acute, equilateral, isosceles, obtuse, right, and scalene. Some answers require just one choice, while others may be answered with two choices.

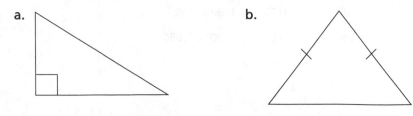

a.   b.

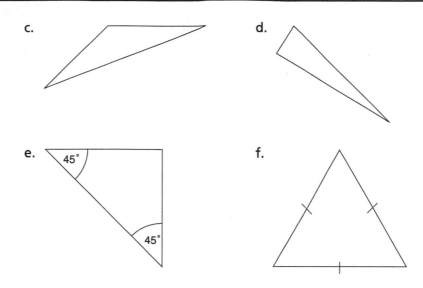

c.

d.

e.

f.

2. How would you describe each of these triangles? Choose from: acute, equilateral, isosceles, obtuse, right, and scalene. Some answers require just one choice, while others may be answered with two choices.

    a. Has three sides of equal length

    b. Contains two angles of equal measure

    c. Contains an angle of 150°

    d. Contains an angle of 90°

3. Find the size of the third angle of each of these triangles.

    a. 30° and 86°

    b. 71° and 48°

    c. 46° and 105°

4. Find the hypotenuse of the following triangle.

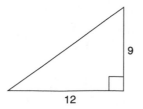

5. Find the hypotenuse of the following triangle.

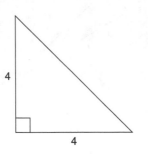

6. Find side *b* of the following triangle.

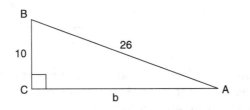

7. △ABC ≅ △A′B′C′. If *a* is 11 and *b* is 19, find the measure of *a′*, *b′*, and *c′*.

8. △DEF ≅ △D′E′F′. If ∠D = 42° and ∠E = 57°, find ∠F′.

9. Fill in the missing side length in the following two congruent triangles.

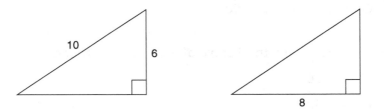

10. Find the height of the following isosceles triangle. Let *a* = 25 and *b* = 20.

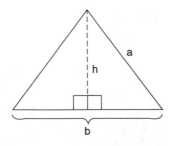

11. △XYZ ~ △GHI. Find the measure of sides *g* and *i*.

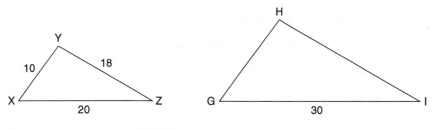

12. Find the perimeter of △FGH.

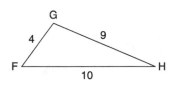

13. Find the perimeter of △DEF if △DEF ~ △ABC.

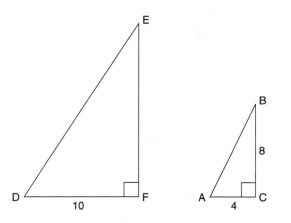

14. Find the area of a triangle with a base of 13 and a height of 9.

15. A right triangle has a height of 10 inches and a hypotenuse of 26 inches. Find its area.

16. How many yards of fencing material are needed to fence in a triangular lot if the lengths of the three sides are 50 feet, 100 feet, and 20 yards?

17. If you walked due south for 30 feet, then walked due west for 40 feet, and then walked directly back to your starting point (that is, on a diagonal), how far did you walk altogether?

18. Find the perimeter and area of a plot of land with the dimensions of a perfect equilateral triangle with a leg of 50 yards.

1. a. right   b. isosceles, acute   c. obtuse   d. acute, scalene

   e. right, isosceles   f. equilateral

2. a. equilateral   b. isosceles   c. obtuse   d. right

3. a. $180° - 160° = 64°$ b. $180° - 119° = 61°$ c. $180° - 151° = 29°$

4. $9^2 + 12^2 = c^2$

   $81 + 144 = c^2$

   $225 = c^2$

   $c = \sqrt{225} = 15$

5. $4^2 + 4^2 = c^2$

   $16 + 16 = c^2$

   $32 = c^2$

   $c = \sqrt{32} = \sqrt{16(2)} = 4\sqrt{2} \approx 5.66$

6. $a^2 + b^2 = c^2$

   $10^2 + b^2 = 26^2$

   $100 + b^2 = 676$

   $b^2 = 576$

   $b = \sqrt{576} = 24$

7. $a' = 11; \; b' = 19$

   $c^2 = 11^2 + 19^2$

   $c^2 = 121 + 361 = 482$

   $c' = \sqrt{482} \approx 21.95$

8. $F' = 81°$

9.

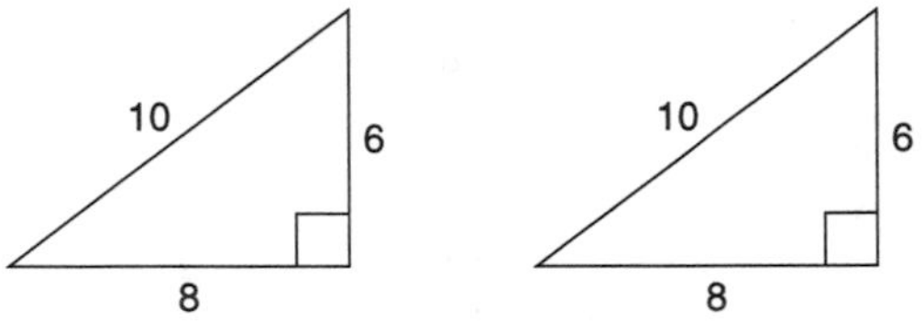

10. $a^2 + \dfrac{1}{2}b^2 = h^2$

    $h^2 + 10^2 = 25^2$

    $h^2 = 625 - 100$

    $h = \sqrt{525} = 5\sqrt{21} \approx 22.91$

11. $\dfrac{30}{20} = \dfrac{3}{2}$

    $g = \dfrac{18}{1}\left(\dfrac{3}{2}\right) = 27$

    $i = \dfrac{10}{1}\left(\dfrac{3}{2}\right) = 15$

12. $4 + 9 + 10 = 23$

13. $\quad 4^2 + 8^2 = c^2$ 14. *Area:*

$\qquad 16 + 64 = c^2 \qquad\qquad\qquad\qquad \dfrac{1}{2}bh = \dfrac{1}{2}(13)(9) = 58.5$

$\qquad 80 = c^2$

$\qquad c = \sqrt{80} = \sqrt{16(5)} = 4\sqrt{5} \approx 8.94$

$\qquad \dfrac{10}{4} = \dfrac{5}{2}$

$\qquad d = \dfrac{5}{2}\left(\dfrac{8}{1}\right) = 20$

$\qquad f = \left(\dfrac{5}{2}\right)\left(4\sqrt{5}\right) = 10\sqrt{5} \approx 22.36$

$\qquad p = 10 + 20 + 10\sqrt{5} = 30 + 10\sqrt{5} \approx 52.36$

15. $a^2 + b^2 = c^2$

$\qquad 10^2 + b^2 = 26^2$

$\qquad 100 + b^2 = 676$

$\qquad b^2 = 576$

$\qquad b = \sqrt{576} = 24$

$\qquad A = \dfrac{1}{2}bh = \dfrac{1}{2}(24)(10) = 120$ square inches

16. $P = 50 + 100 + 60 = 210$ feet $= 70$ yards

17. $a^2 + b^2 = c^2$

$\qquad 30^2 + 40^2 = c^2$

$\qquad 900 + 1{,}600 = c^2$

$\qquad 2{,}500 = c^2$

$\qquad c = \sqrt{2{,}500} = 50$

The distance walked is $30 + 40 + 50 = 120$ feet.

18. Perimeter $= 50 + 50 + 50 = 150$ yards

$\qquad h^2 + 25^2 = 50^2$

$\qquad h = \sqrt{2{,}500 - 625} = 25\sqrt{3}$

$\qquad$Area $= \dfrac{1}{2}(50)(25\sqrt{3}) = 625\sqrt{3} \approx 1{,}082.53$ square yards

How did you do? On rare occasions readers *do* get every problem right. If you happen to be one of these people, then you have our permission to skip the first four sections of this chapter and go directly to the "Applications" section. If you had trouble with several of the problems in this pretest, you definitely need to work your way through the entire chapter. You may skip the next paragraph and go directly to the "Types of Triangles" section.

If you're reading this, then you must be somewhere in the middle. If you got the answers to questions 1, 2, and 3 entirely right, you may skip the section. If you got questions 4, 5, and 6 entirely right, you may skip the section on the Pythagorean theorem. If you answered questions 7 to 11 correctly, you may skip the "Congruent and Similar Triangles" section. And if you got questions 12 to 14 right, you may skip the "Perimeter and Area of a Triangle" section.

Types of Triangles

We're going to build on much of what we covered in the "Line Segment" section of chapter 1, now using the angles we defined there to define triangles. The first three triangles that we'll cover—acute, obtuse, and right triangles—should be very familiar.

An acute triangle contains three angles of less than 90°. Each of the angles contained in the following triangle is less than 90°.

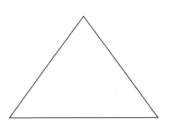

An obtuse triangle contains one angle greater than 90°. As you can see, the following triangle has one angle on top that is greater than 90°.

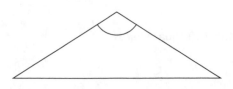

A *right triangle contains a right angle.* The following triangle contains a right angle. The box shown in the following figure represents a right angle. A right angle is 90°.

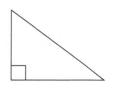

Next we introduce three more types of triangles. *An equilateral triangle has sides of equal lengths, and all three angles are 60°.* The following triangle is an equilateral triangle. The single line drawn across each side of the following triangle indicates that the sides are of equal length.

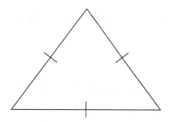

An isosceles triangle has two sides of equal length, and the angles opposite the two equal sides have equal measurements. The following triangle is an isosceles triangle.

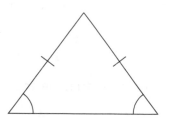

A scalene triangle has three sides of different lengths, and all three angles have different measurements. The following triangle is a scalene triangle.

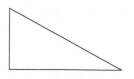

Do you recall the formula $(n - 2)180° = (3 - 2)180° = (1)180° = 180°$? We'll assume that the sum of the angles of any triangle is 180°.

Example 1:

If a triangle contains angles of 29° and 58°, how much is the measure of the third angle?

Solution:

We'll write an equation to show the sum of the angles is 180°. We'll let x represent the measure of the unknown angle and solve for x.

$180° = 29° + 58° + x°$	Add 29 and 58.
$180° = 87° + x°$	Subtract 87 from both sides of the equation.
$93° = x°$	

Example 2:

Find the measure of the third angle of an isosceles triangle if its two base angles are 50°.

Solution:

Again we will write and solve an equation to represent the sum of the angles of the triangle. We already know the base angles of an isosceles triangle are congruent; in this case, both are 50°.

$180° = 50° + 50° + x°$	Add 50 and 50.
$180° = 100° + x°$	Subtract 100 from both sides of the equation.
$80° = x°$	

Example 3:

Find the measure of the base angles of an isosceles triangle the third angle of which measures 56°.

Solution:

Again we'll write and solve an equation to represent the sum of the measures of the angles of a triangle. This time we're asked to solve for the measure of the base angles of an isosceles triangle, which we can assume are equal.

$180° = x° + x° + 56°$	Combine the $x°$ with the $x°$.
$180° = 2x° + 56°$	Subtract 56° from both sides of the equation.
$124° = 2x°$	Divide both sides of the equation by 2.
$62° = x$	

Example 4:

Find the measure of the angles of an equilateral triangle.

Solution:

All the angles of an equilateral triangle are congruent, or equal.

$180° = x° + x° + x°$ Combine the $x°$'s.

$180° = 3x°$ Divide both sides of the equation by 3.

$60° = x°$

   And now for two other bits of wisdom that should be pretty obvious. First, the length of one side of a triangle is always less than the sum of the other two sides. Second, in a triangle, the largest side is opposite the largest angle, the smallest side is opposite the smallest angle, and the middle-length side (if there is one), is opposite the middle-size angle.

SELF-TEST 1

1. Describe each of these triangles.

a.

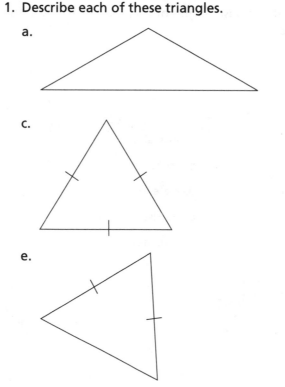

b.

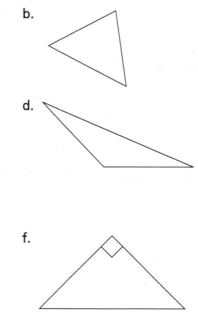

c.

d.

e.

f.

2. How would you describe each of these triangles?

   a. Contains a right angle

   b. Has two sides of equal length

   c. Contains three angles of less than 90°

   d. Has three angles of different measurements and three sides of different length

   e. Contains one angle greater than 90°

   f. Contains three angles of equal measure

3. Find the measure of the third angle of each of these triangles.

   a. The first angle is 40°, the second is 75°.

   b. The first angle is 95°, the second is 36°.

   c. The first angle is 48°, the second is 107°.

4. Find the measure of the third angle of an isosceles triangle the base angles of which measure 45°.

5. Find the measure of the base angles of an isosceles triangle the third angle of which measures 68°.

ANSWERS

1. a. obtuse and/or scalene

   b. acute

   c. equilateral

   d. scalene and/or obtuse

   e. isosceles

   f. right

2. a. right

   b. isosceles

   c. acute

   d. scalene

   e. obtuse

   f. equilateral

3. a. $180° = 40° + 75° + x°$ Add 40° and 75°.

      $180° = 115° + x°$ Subtract 115° from both sides of the equation.

      $65° = x°$

   b. $180° = 95° + 36° + x°$ Add 95° and 36°.

      $180° = 131° + x°$ Subtract 131° from both sides of the equation.

      $49° = x°$

   c. $180° = 48° + 107° + x°$ Add 48° and 107°.

      $180° = 155° + x°$ Subtract 155° from both sides of the equation.

      $25° = x°$

4. $180° = 45° + 45° + x°$ Add 45° and 45°.

   $180° = 90° + x°$ Subtract 90° from both sides of the equation.

   $90° = x°$

5. $180° = 68° + 2x°$ Subtract 68° from both sides of the equation.

   $112° = 2x°$ Divide both sides of the equation by 2.

   $56° = x°$

Pythagorean Theorem

Complete this equation: $a^2 + b^2 = $? Did you say c^2? Know what this equation is called? *It's the Pythagorean theorem: the square of the hypotenuse of a right triangle is equal to the sum of the squares of its two other sides.* Keep in mind that the Pythagorean theorem applies only to right triangles. The following right triangle shows sides a, b, and c. Each side is opposite its corresponding angle. That is, side a is opposite angle A, side b is opposite angle B, and side c is opposite angle C. Usually the right angle is labeled angle C. The side opposite the right angle is called the hypotenuse.

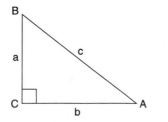

One of the great things about the Pythagorean theorem is that it enables us to do an endless variety of $a^2 + b^2 = c^2$ applications.

Example 5:

Find the measure of the hypotenuse in the following triangle.

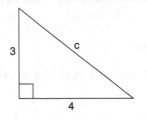

Solution:

This is a right triangle, so we'll use the Pythagorean theorem.

$a^2 + b^2 = c^2$	Substitute the values for sides a, b, and c into the equation.
$3^2 + 4^2 = c^2$	Square the numbers.
$9 + 16 = c^2$	Add the numbers.
$25 = c^2$	Take the square root of both sides of the numbers.
$5 = c$	

Sometimes you'll want to find the measure of one of the legs of the triangle instead of the measure of a triangle's hypotenuse.

Example 6:

Find the measure of side b in the following triangle.

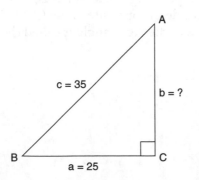

Solution:

To find the measure of side *b*, we'll use the Pythagorean theorem.

$a^2 + b^2 = c^2$ Substitute 25 for *a* and 35 for *c*.

$25^2 + b^2 = 35^2$ Square both numbers.

$625 + b^2 = 1{,}225$ Subtract 625 from both sides of the equation.

$b^2 = 600$ Take the square root of both sides of the

$b = \sqrt{600} = \sqrt{100(6)} = 10\sqrt{6} \approx 24.49$ equation.

The symbol $\approx$ means "approximately" and is used to show that a value has been rounded off in decimal form. The measure of side b is *exactly* $10\sqrt{6}$, which is *approximately* 24.49.

Example 7:

Find the measure of side a.

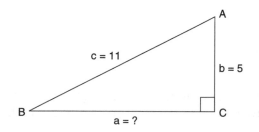

Solution:

$a^2 + b^2 = c^2$ Substitute 11 for *c*, 5 for *b*.

$a^2 + 25 = 121$ Square the numbers.

$a^2 = 96$ Subtract 25 from both sides of the equation.

$a = \sqrt{96} = \sqrt{16(6)} = 4\sqrt{6} \approx 9.8$

Example 8:

Find the height of the following isosceles triangle. Keep in mind that the height, *h*, of an isosceles triangle bisects the base. Let *a* = 15 and *b* = 10.

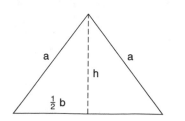

Solution:

Since the height of an isosceles triangle bisects the base, we can assume it forms two back-to-back right triangles. We'll use the Pythagorean theorem to find the measure of the height. Side a is opposite the right angle, so it's the hypotenuse of the triangle. When we fill in a value for $\frac{1}{2}b$, we'll use $\frac{1}{2}\left(\frac{10}{1}\right) = 5$.

$$h^2 + \left(\frac{1}{2}b\right)^2 = a^2$$

$$h^2 = a^2 - \left(\frac{1}{2}b\right)^2 \qquad\qquad \text{Substitute 15 for } a \text{ and 5 for } \frac{1}{2}b.$$

$$h^2 = 15^2 - 5^2 \qquad\qquad \text{Square both numbers.}$$

$$h^2 = 225 - 25 \qquad\qquad \text{Subtract.}$$

$$h^2 = 200 \qquad\qquad\qquad \text{Take the square root of both sides}$$

$$h = \sqrt{200} = \sqrt{100(2)} = 10\sqrt{2} \approx 14.14 \qquad \text{of the equation.}$$

SELF-TEST 2

When you find the measure of the sides in the next triangles, state the exact value of your answer and then approximate your answer to two decimal places.

1. Find the measure of the hypotenuse of the following triangle.

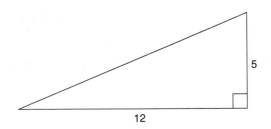

5

12

2. Find the measure of the hypotenuse of the following triangle.

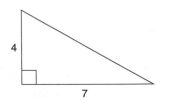

4

7

3. Find the measure of side *a* of the following triangle.

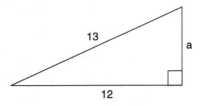

4. Find the measure of side *b* of the following triangle.

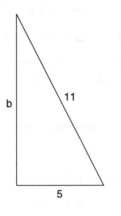

5. Find the measure of sides *a* of the following triangle.

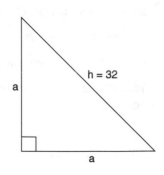

1. $a^2 + b^2 = c^2$ Substitute 5 for a and 12 for b.

   $5^2 + 12^2 = c^2$ Square both numbers.

   $25 + 144 = c^2$ Add the numbers.

   $169 = c^2$ Take the square root of both sides of the equation.

   $13 = c$

2. $c^2 = a^2 + b^2$ Substitute 4 for a and 7 for b.

   $c^2 = 4^2 + 7^2$ Square both numbers.

   $c^2 = 16 + 49$ Add the numbers.

   $c^2 = 65$ Take the square root of both sides of the equation.

   $c = \sqrt{65} \approx 8.06$

3. $a^2 + b^2 = c^2$ Substitute 12 for b and 13 for c.

   $a^2 + 12^2 = 13^2$ Square both numbers.

   $a^2 + 144 = 169$ Subtract 144 from both sides of the equation.

   $a^2 = 25$ Take the square root of both sides of the equation.

   $a = 5$

4. $a^2 + b^2 = c^2$ Substitute 5 for a and 11 for c.

   $5^2 + b^2 = 11^2$ Square both numbers.

   $25 + b^2 = 121$ Subtract 25 from both sides of the equation.

   $b^2 = 96$ Take the square root of both sides of the equation.

   $b = \sqrt{96} = 4\sqrt{6} \approx 9.8$

5. $a^2 + a^2 = h^2$ Combine a^2 and a^2.

   $2a^2 = (32)^2$ Square 32.

   $2a^2 = 1,024$ Divide both sides of the equation by 2.

   $a^2 = 512$ Take the square root of both sides of the equation.

   $a = \sqrt{512} = \sqrt{256(2)} = 16\sqrt{2} \approx 22.63$

Congruent and Similar Triangles

Congruent polygons are exact duplicates of each other: They have the same size and shape. Therefore, *congruent triangles have the same size and shape.* In the figure nearby, triangles RST and R'S'T' are congruent. That means the corresponding sides and angles of the triangles have equal measure. Two triangles are congruent if there is a one-to-one correspondence with respect to:

a. two sides and the included angle, or

b. two angles and the included side, or

c. three sides, or

d. a right angle, the hypotenuse, and one other side.

The included angle is the angle between the sides. The included side is the side between the angles. For example, in the following triangles, the *included angle* between sides RS and RT is angle R. The included side between angles R and T is side RT.

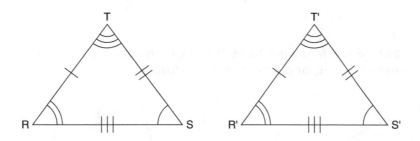

The symbol for congruency is ≅, which simply means "is congruent to." Symbolically, then, RST ≅ R'S'T'. The two triangles are congruent because their corresponding sides and angles are congruent. Look at the sides of these triangles: RS ≅ R'S', ST ≅ S'T', and RT ≅ R'T'. Just as the sides of the triangles are congruent, you'll find that their angles are congruent. ∠R ≅ ∠R'; ∠S ≅ S'; ∠T ≅ T'. Ready to apply some of your knowledge about congruency?

Example 9:

Which two of the following three triangles are congruent? State which property from above, a, b, c, or d, led to your conclusion.

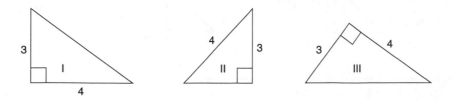

Solution:

Triangles I and III; the reason is property a, two sides and the included angle.
   Try another one.

Example 10:

Which two of the following three triangles are congruent? State which property from page 41, a, b, c, or d, led to your conclusion.

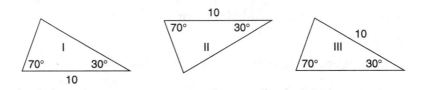

Solution:

Triangles I and II; the reason is property b, the angles and the included side.

Example 11:

Which two of the following three triangles are congruent? State which property from page 41, a, b, c, or d, led to your conclusion.

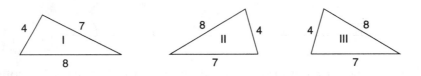

Solution:

Triangles II and III; the reason is property c, three sides.

Example 12:

Find the measure of the missing sides and angles in the following two congruent triangles.

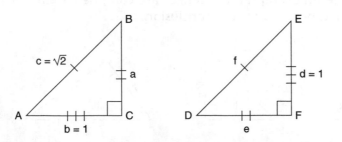

Solution:

To find the measure of side *a* of ΔABC, we'll use the Pythagorean theorem.

$a^2 + b^2 = c^2$ Substitute $\sqrt{2}$ for *c* and 1 for *b*.

$a^2 + 1^2 = (\sqrt{2})$ Square both numbers. $(\sqrt{2})^2 = 2$

$a^2 + 1 = 2$ Subtract 1 from each side.

$a^2 = 1$ Take the square root of both sides of the equation.

$a = 1$

A much faster and easier way to find the measure of side *a* is to use our knowledge of congruent triangles and just say the measure of side *d,* which is given, is congruent to the measure of side *b*.

Now that we know the measure of the sides of ΔABC, we also know the measure of the sides of ΔDEF. Side *d* = 1, side *f* = $\sqrt{2}$, and side *e* = 1. If we know that angle A is 45°, then we know that angle D is also 45°. Angles C and F are right angles, so we know they each measure 90°. To find the measure of angles B and E, we use our knowledge that the sum of the angles of a triangle is 180°. The measure of angles B and E is 180° − 90° − 45° = 45°. Did you notice that these triangles have two angles and two sides of equal measure? What type of triangle is this? If you said isosceles, you're correct.

Moving right along, let's now consider similar triangles. *Triangles are similar if their corresponding angles are congruent and their sides are in proportion.* So similar triangles have the same shape, but not the same size. If two triangles have the same size and shape, they are congruent. We use the symbol ~ to denote similarity. For example, because the following triangles are similar, we write ΔDEF ~ ΔD′E′F′.

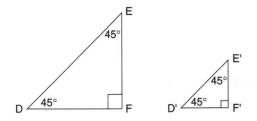

The corresponding sides of similar triangles are opposite congruent angles. For convenience sake, corresponding sides and angles are usually but not always identified by the same letters with primes. Thus, in the two triangles ΔFGH and ΔF′G′H′, ΔFGH ~ ΔF′G′H′, since ∠F = 90° = ∠F′; ∠G = 53° = ∠G′; ∠H = 37° = ∠H′; and $\dfrac{f}{f'} = \dfrac{g}{g'} = \dfrac{h}{h'}$, or $\dfrac{10}{5} = \dfrac{8}{4} = \dfrac{6}{3}$. The angles are equal and the sides have the same ratio.

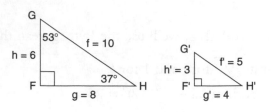

Example 13:

We know that, by definition, the corresponding sides of similar triangles are in proportion. $\triangle ABC \sim \triangle A'B'C'$. See if you can figure out the length of the sides of $\triangle A'B'C'$.

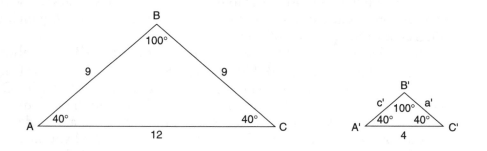

Solution:

The ratio from b to b' is $\dfrac{b}{b'} = \dfrac{12}{4} = \dfrac{3}{1}$. $\triangle A'B'C'$ is one-third the size of $\triangle ABC$. Therefore, all we have to do to find the measure of sides c' and a' is divide sides c and a by 3 or multiply by $\dfrac{1}{3}$. $c' = \dfrac{9}{3} = 3$, $a' = \dfrac{9}{3} = 3$.

SELF-TEST 3

1. $\triangle ABC \cong \triangle A'B'C'$. If a is 12 and b is 9, find a' and b'.

2. $\triangle DEF \cong \triangle D'E'F'$. If $\angle D = 35°$ and $\angle E = 65°$, find $\angle D'$ and $\angle E'$.

3. $\triangle CDE \cong \triangle C'D'E'$. If $\angle D = 48°$ and $\angle E = 17°$, find $\angle F'$.

4. Fill in the missing lengths in the following two sets of congruent triangles.

   a.

b.

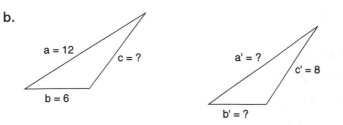

5. Find the height of the following isosceles triangle. Let $a = 20$ and $b = 18$.

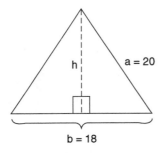

6. Find the length of the leg of an isosceles right triangle the height of which is 18 and the base of which is 22.

7. $\triangle ABC \sim \triangle GHI$. Find the lengths of the sides g and h of $\triangle GHI$.

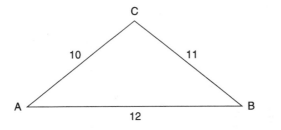
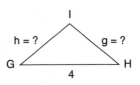

8. Find the lengths of c' and a'.

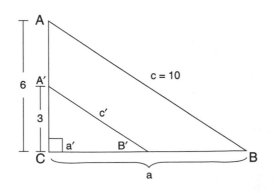

ANSWERS

1. $a' = 12$, $b' = 9$

2. $\angle D' = 35°$, $\angle E' = 65°$

3. $F' = 180° - 48° - 17° = 115°$

4. a. $a' = 9$, $c' = 7$, $b = 8$

   b. $a' = 12$, $b' = 6$, $c = 8$

5. $h^2 + \left(\dfrac{1}{2}b\right)^2 = a^2$

   $h^2 = a^2 - \left(\dfrac{1}{2}b\right)^2$

   $h^2 = 20^2 - 9^2$

   $h^2 = 400 - 81$

   $h^2 = 319$

   $h = \sqrt{319} \approx 17.86$

6. $a^2 = 18^2 + 11^2$

   $a^2 = 324 + 121$

   $a^2 = 445$

   $a = \sqrt{445} \approx 21.1$

7. $\dfrac{4}{12} = \dfrac{1}{3}$

   $h = \dfrac{1}{3}\left(\dfrac{10}{1}\right) = \dfrac{10}{3} = 3\dfrac{1}{3}$

   $g = \dfrac{1}{3}\left(\dfrac{11}{1}\right) = \dfrac{11}{3} = 3\dfrac{2}{3}$

8. $\dfrac{3}{6} = \dfrac{1}{2}$

   $c' = \dfrac{1}{2}\left(\dfrac{10}{1}\right) = 5$

   $a^2 + b^2 = c^2$

   $a^2 = c^2 - b^2$

   $a^2 = 100 - 36$

   $a = \sqrt{64} = 8$

   $a' = \dfrac{1}{2}\left(\dfrac{8}{1}\right) = 4$

Perimeter and Area of a Triangle

Perimeter

We are going to start with a very simple definition. *The perimeter of a triangle is the sum of its sides.*

Example 14:

What is the perimeter of the following triangle?

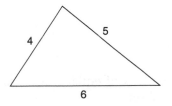

Solution:

P = 4 + 5 + 6 = 15

As Ross Perot used to say, "It's as simple as that."

Example 15:

See if you can find the perimeter of the following right triangle.

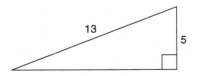

Solution:

First we have to use the Pythagorean theorem to find the measure of the missing side.

$$a^2 + b^2 = c^2$$
$$5^2 + b^2 = 13^2$$
$$25 + b^2 = 169$$
$$b^2 = 169 - 25$$
$$b^2 = 144$$
$$b = 12$$

So P = 13 + 5 + 12 = 30.

Now let's move on to isosceles triangles.

Example 16:

Find the perimeter of this isosceles triangle.

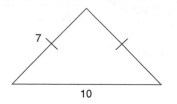

Solution:

Remember that an isosceles triangle has two legs of equal length, so if you know the length of one of the equal legs, you're all set.

P = 7 + 7 + 10 = 24

Example 17:

Find the perimeter of this triangle. (Hint: What kind of triangle is this?)

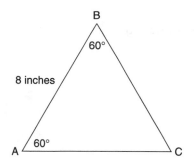

Solution:

This is an equilateral triangle: If ∠A and ∠B are each 60°, then ∠C must also be 60°. Because one side is 8 inches, each of the other sides must also be 8 inches. 3 × (8 inches) = 24 inches.

Area

The area of a triangle equals one-half the product of a side and the altitude to that side. Put more simply, the area of a triangle is equal to $\frac{1}{2}$ its base times its height, or $A = \frac{1}{2} bh$.

Example 18:

Find the area of the following triangle.

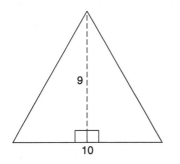

Solution:

$$A = \frac{1}{2}bh = \frac{1}{2}\left(\frac{10}{1}\right)\left(\frac{9}{1}\right) = 45$$

Example 19:

Find the area of the following right triangle.

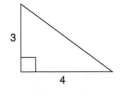

Solution:

$$A = \frac{1}{2}bh = \frac{1}{2}\left(\frac{4}{1}\right)\left(\frac{3}{1}\right) = 6$$

Example 20:

Now find the area of the following equilateral triangle. (Trust us, you have enough information to solve this problem.)

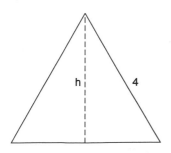

Solution:

Because it's an equilateral triangle, we can assume all sides are have a length of 4 units. Half the base is 2. We'll use the Pythagorean theorem to find the height.

$$2^2 + h^2 = 4^2$$

$$4 + h^2 = 16$$

$$h^2 = 12$$

$$h = \sqrt{12} = \sqrt{4(3)} = 2\sqrt{3}$$

$$A = \frac{1}{2}bh = \frac{1}{2}\left(\frac{2}{1}\right)\left(\frac{2\sqrt{3}}{1}\right) = 2\sqrt{3} \approx 3.46$$

Example 21:

Find the area of $\triangle A'B'C'$ using the following similar triangles.

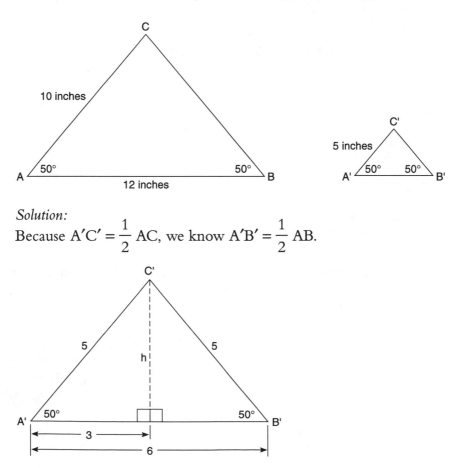

Solution:

Because $A'C' = \dfrac{1}{2}\,AC$, we know $A'B' = \dfrac{1}{2}\,AB$.

$$3^2 + h^2 = 5^2$$
$$h^2 = 25 - 9$$
$$h = \sqrt{16} = 4$$
$$A = \frac{1}{2}bh = \frac{1}{2}\left(\frac{3}{1}\right)\left(\frac{4}{1}\right) = 6 \text{ square inches}$$

SELF-TEST 4

1. Find the perimeter of a triangle with sides 4, 8, and 9.

2. Find the perimeter of the following triangle.

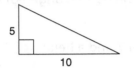

3. Find the perimeter of ΔA′B′C′ if ΔABC ~ ΔA′B′C′.

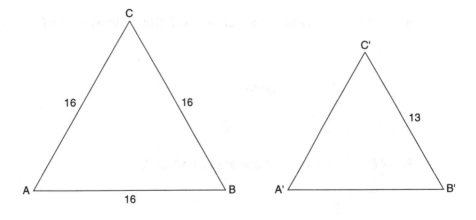

4. Find the perimeter of ΔF′G′H′ if ΔF′G′H′ ~ ΔFGH.

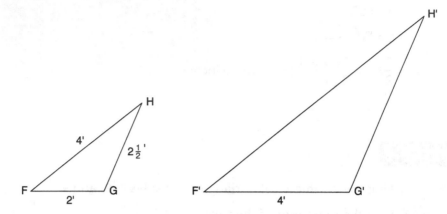

5. What is the perimeter of a triangle containing two angles of 60° and a leg 4 inches in length?

6. Find the area of a triangle with a base of 7 and a height of 8.

7. A right triangle has a leg that is 5 inches long and a hypotenuse of 13 inches. Find its area.

8. Find the perimeter and area of ΔRST. (Hint: What kind of triangle is this?)

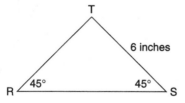

9. ΔABC ≅ ΔA′B′C′. Find the area of ΔA′B′C′.

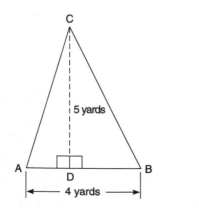
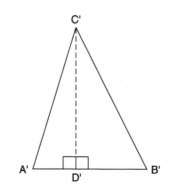

10. $\triangle CDE \sim \triangle C'D'E'$. Find the perimeter and area of $\triangle C'D'E'$.

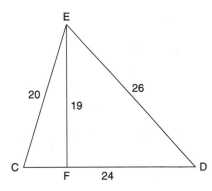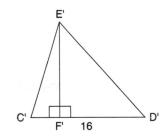

11. Find the area of $\triangle ABE$.

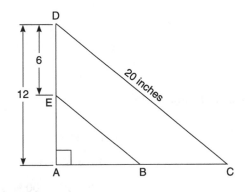

1. $P = 4 + 8 + 9 = 21$

2. $a^2 + b^2 = c^2$

   $5^2 + 10^2 = c^2$

   $25 + 100 = c^2$

   $125 = c^2$

   $c = \sqrt{125} = \sqrt{25(5)} = 5\sqrt{5} \approx 11.18$

   $P = 5 + 10 + 5\sqrt{5} \approx 26.18$

3. $P = 13 + 13 + 13 = 39$

4. $\dfrac{4}{2} = 2$

   $4(2) = 8$

   $\left(2\dfrac{1}{2}\right)(2) = 5$

   $P = 4 + 8 + 5 = 17$

5. This is an equilateral triangle. The third angle must also be 60°. Because the three legs of an equilateral triangle are, by definition, equal, $P = 4 + 4 + 4 = 12$.

6. $A = \frac{1}{2}bh = \frac{1}{2}\left(\frac{7}{1}\right)\left(\frac{8}{1}\right) = 28$

7. $a^2 + b^2 = c^2$

   $a^2 + 5^2 = 13^2$

   $a^2 = 169 - 25$

   $a = \sqrt{144} = 12$

   $A = \frac{1}{2}bh = \frac{1}{2}\left(\frac{5}{1}\right)\left(\frac{12}{1}\right) = 30$ square inches

8. This is a right triangle. $180° - 45° - 45° = 90°$

   $a^2 + b^2 = c^2$

   $6^2 + 6^2 = c^2$

   $36 + 36 = c^2$

   $72 = c^2$

   $c = \sqrt{72} = \sqrt{36(2)} = 6\sqrt{2} \approx 8.49$

   $P = 6 + 6 + 6\sqrt{2} \approx 20.49$

   $A = \frac{1}{2}bh = \frac{1}{2}\left(\frac{6}{1}\right)\left(\frac{6}{1}\right) = 18$ square inches

9. $A = \frac{1}{2}\left(\frac{4}{1}\right)\left(\frac{5}{1}\right) = 10$ square yards

10. $\frac{16}{24} = \frac{2}{3}$

    $\frac{20}{1}\left(\frac{2}{3}\right) = \frac{40}{3} = 13\frac{1}{3}$

    $\frac{26}{1}\left(\frac{2}{3}\right) = \frac{52}{3} = 17\frac{1}{3}$

    $P = 16 + 13\frac{1}{3} + 17\frac{1}{3} = 46\frac{2}{3}$

    $h = \left(\frac{2}{3}\right)\left(\frac{19}{1}\right) = \frac{38}{3}$

    Area of $\triangle C'D'E' = \frac{1}{2}\left(\frac{16}{1}\right)\left(\frac{38}{3}\right) = \frac{304}{3} = 101\frac{1}{3}$

    Area of $\triangle CDE = \left(\frac{1}{2}\right)\left(\frac{24}{1}\right)\left(\frac{19}{1}\right) = 228$

11. $a^2 + b^2 = c^2$

    $12^2 + b^2 = 20^2$

    $144 + b^2 = 400$

    $b^2 = 400 - 144$

    $b = \sqrt{256} = 16$

    $AC = 16, EB = \frac{20}{2} = 10, AB = \frac{16}{2} = 8$

    Area $= \frac{1}{2}\left(\frac{8}{1}\right)\left(\frac{6}{1}\right) = 24$ square inches

Applications

Our students often ask, "How is this material relevant in the real world?" Or "How does any of this affect our daily lives?" In this section, and in the concluding sections of the next few chapters, we'll show you how to apply the skills you've learned to everyday situations. A few of these applications are a bit far-fetched. But hey, so is life sometimes.

Example 22:

How many feet of fencing material are needed to fence in a triangular garden if its three sides are 12 feet, 9 feet, and 8 feet long?

Solution:

$12 + 9 + 8 = 29$ feet

Example 23:

If you walk due north for 3 miles, then walk due east for 4 miles, then walk directly back to your starting point (that is, on a diagonal), how far would you walk altogether?

Solution:

Sometimes it helps to draw the problem first.

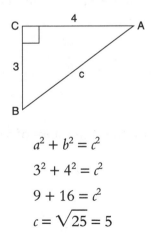

$$a^2 + b^2 = c^2$$
$$3^2 + 4^2 = c^2$$
$$9 + 16 = c^2$$
$$c = \sqrt{25} = 5$$

The distance walked is $3 + 4 + 5 = 12$ miles.

Example 24:

The course of the Flatbush Frolics Fun Run is 1,200 yards in one direction, then a sharp turn 60° to the left. The course continues straight for another 1,200 yards until the runners come to another sharp 60° turn to the left, and then they run straight to the finish line, which is exactly where they started the race. How far did they run?

Solution:

When you draw the course, the solution becomes very obvious.

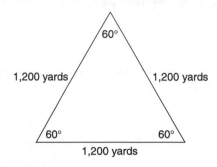

They ran $1,200(3) = 3,600$ yards.

Example 25:

If sod costs $2 per square foot, how much would it cost to cover a triangular plot of land with a height of 7 feet and a base of 10 feet?

Solution:

First we have to find the area of the triangle. Then we'll multiply the area by the cost per square foot.

$$A = \frac{1}{2}bh = \frac{1}{2}\left(\frac{10}{1}\right)\left(\frac{7}{1}\right) = 35 \text{ square feet}$$

$\text{Cost} = 35(\$2) = \$70$

Example 26:

Find the perimeter and area of a plot of land with the dimensions of a perfect equilateral triangle with a leg of 40 feet.

Solution:

$P = 40 + 40 + 40 = 120$ feet

$h^2 + 20^2 = 40^2$

$h^2 = 1,600 - 400$

$h^2 = 1,200$

$h = \sqrt{1,200} = \sqrt{400(3)} = 20\sqrt{3} \approx 34.64$

$$A = \frac{1}{2}\left(\frac{40}{1}\right)\left(\frac{20\sqrt{3}}{1}\right) = 400\sqrt{3} \approx 692.82$$

Example 27:

A tree casts a 15-foot shadow at a time when a nearby upright pole that is 6 feet high casts a 2-foot shadow. Find the height of the tree if both the tree and the pole make right angles with the ground.

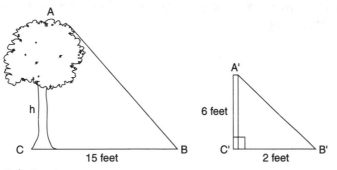

Solution:

Because the pole and the tree are so close to each other, the rays of the sun strike the ground at congruent angles, so $\angle B \cong \angle B'$. And since the tree and pole make right angles with the ground, $\angle C \cong \angle C'$. Therefore, $\triangle ABC \cong \triangle A'B'C'$, $\dfrac{h}{6} = \dfrac{15}{2}$, $2h = 90$, and $h = 45$ feet.

Example 28:

A car traveled due west for 60 miles, then traveled due south for 80 miles and returned directly (on a diagonal) to its starting point. If the car gets 12 miles per gallon, how many gallons of gas are needed for the trip?

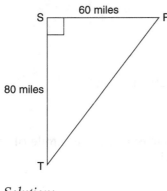

Solution:

$a^2 + b^2 = c^2$

$60^2 + 80^2 = c^2$

$3,600 + 6,400 = c^2$

$10,000 = c^2$

$c = 100$

The distance traveled is 240 miles. You need $240 \div 12 = 20$ gallons of gas to make the trip.

Example 29:

A guy wire attached to a 200-foot radio tower is fastened to the ground 50 feet from the tower. How long is the guy wire?

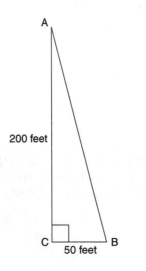

Solution:

$a^2 + b^2 = c^2$

$50^2 + 200^2 = c^2$

$2{,}500 + 40{,}000 = c^2$

$42{,}500 = c^2$

$c = \sqrt{42{,}500} = \sqrt{2{,}500(17)} = 50\sqrt{17} \approx 206.16$ feet

Example 30:

If it takes a farmer an hour to plow one-quarter of a square mile of land, how long does it take her to plow the following field?

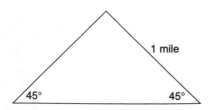

Solution:

We know this is an isosceles triangle because the base angles are congruent. Therefore, we know that the side opposite the 45° is 1 mile.

The area $= \frac{1}{2}(1)(1) = \frac{1}{2}$ square miles. One-half is 2 times one-quarter. Therefore, if it takes 1 hour to plow $\frac{1}{4}$ square mile, it would take twice that, or 2 hours, to plow $\frac{1}{2}$ square mile.

Now it's time to raise the level of the problem a little. The next few problems assume you know basic algebra.

Example 13:

Imagine a lot in the shape of a right triangle. The shorter leg of the triangle is 1 meter less than the longer leg. The hypotenuse is 1 meter more than the longer leg. Find the perimeter of the lot.

Solution:

We'll start by drawing this one. The length of the hypotenuse and the shorter leg is based on the longer leg, so we'll begin by letting $x =$ the length of the longer leg. Then $x - 1$ equals the length of the shorter leg and $x + 1$ equals the length of the hypotenuse.

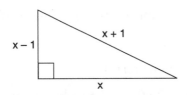

We'll use the Pythagorean theorem to write an equation to represent the relationship between the sides of the triangle.

$x^2 + (x - 1)^2 = (x + 1)^2$	Write the binominals twice and distribute.
$x^2 + (x - 1)(x - 1) = (x + 1)(x + 1)$	Distribute.
$x^2 + x^2 - x - x + 1 = x^2 + x + x + 1$	Combine like terms.
$2x^2 - 2x + 1 = x^2 + 2x + 1$	Move all terms to the left side of the equation.
$2x^2 - 2x + 1 - x^2 - 2x - 1 = 0$	Combine like terms.
$x^2 - 4x = 0$	Factor out the common factor of x.
$x(x - 4) = 0$	Set the factors equal to 0 and solve for x.
$x = 0, x - 4 = 0, x = 4.$	0 can't be the measure of a side of the triangle.

The measures of the sides are 4, 3, and 5. $P = 4 + 3 + 5 = 12$

Example 32:

A kite is flying on a 50-foot string. How high is it above the ground if its height is 10 feet more than the horizontal distance from the person flying it? Assume the string is being held at ground level and is let out all the way.

Solution:

We'll start by letting x equal the height of the kite. Then the horizontal distance from the person flying it is $x - 10$.

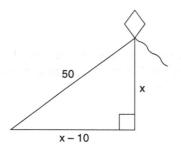

We'll use the Pythagorean theorem to write this as an equation.

$50^2 = x^2 + (x - 10)^2$	Write the binominal twice and distribute.
$2{,}500 = x^2 + (x - 10)(x - 10)$	
$2{,}500 = x^2 + x^2 - 10x - 10x + 100$	Combine like terms.
$2{,}500 = 2x^2 - 20x + 100$	Move the 2,500 to the right side of the equation.
$0 = 2x^2 - 20x - 2{,}400$	Factor and solve for x.
$0 = 2(x^2 - 10x - 1{,}200)$	
$0 = 2(x - 40)(x + 30)$	
$x = 40, x = -30$	x can't be negative, so $x = -30$ is not acceptable.

The horizontal distance, $x - 10$, is $40 - 10$, or 30 feet. The height is 40 feet.

Example 33:

An airplane is approaching Newark Airport at an altitude of 3,567 feet. If the horizontal distance from the plane to the runway is 1.5 miles, find the diagonal distance from the plane to the runway.

Solution:

We'll begin by drawing a triangle to represent the situation.

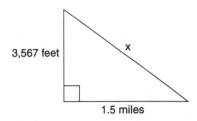

Next we'll use the Pythagorean theorem to solve the problem. Because one of the measurements is in feet and the other is in miles, we must convert to one measurement, in this case, feet. 1.5 miles = (1.5)(5,280 feet) = 7,920 feet.

$$x^2 = (3,567)^2 + (7,920)^2$$
$$x^2 = 75,449,889$$
$$x = \sqrt{75,449,889} = 3\sqrt{8,383,321} \approx 8,686.19 \text{ feet}$$

The distance from the plane to the runway is 8,686.9 feet, or approximately 1.65 miles.

SELF-TEST 5

1. How much would it cost to fence in a triangular lot if the lengths of its three sides are 20 feet, 23 feet, and 29 feet and the cost per foot is $15?

2. Find the perimeter and area of the vacant lot represented by $\triangle$FGH. If fencing costs $58 per yard, how much would it cost to fence in the lot? If sod costs $14 per square yard, how much would it cost to cover the lot with sod?

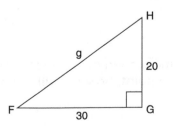

3. If you walked 8 miles in one direction, made a 90° right turn, walked another 8 miles, then made a 45° right turn and returned directly to your starting point, how far would you need to walk to get back to where you started?

4. If artificial turf costs $10 a square yard, how much would it cost to cover a triangular field that had a height of 12 yards and a base of 22 yards?

5. A dermatologist doing a hair transplant knows he needs 28 hair plugs to cover each square inch of baldness. If he needs to cover a bald spot with the dimensions of a perfect equilateral triangle with a leg of 2 inches, how many hair plugs will he need?

6. According to an old pirate's map, a treasure chest is buried exactly 150 paces to the northwest of a very prominent tree. Unfortunately, a thick stand of trees has since grown between that tree and the buried treasure. To find the treasure, you could start pacing from the tree to the west, then make a 90° right turn and pace to the north. To find the exact location, how far would you pace to the north and then to the west? Hint: You would pace the same distance to the north as to the west, so you need to find that distance. Start by drawing a map.

7. $\triangle CDE \sim \triangle C'D'E'$. $\triangle CDE$ is a map drawn on a scale of 1 inch to a corresponding distance of 3 miles on $\triangle C'D'E'$. How much is the distance between C' and D'? What is the perimeter of $\triangle C'D'E'$?

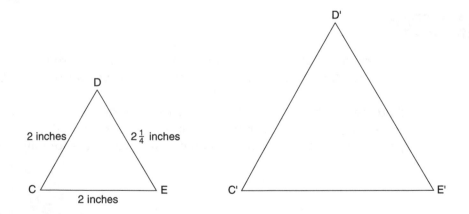

8. If a plane travels due east for 150 miles, then goes due south for 200 miles, then flies directly back to its starting point, how far will the plane have traveled?

9. If carpeting costs $10 per square yard, how much would it cost to carpet the two rooms represented by the following triangles?

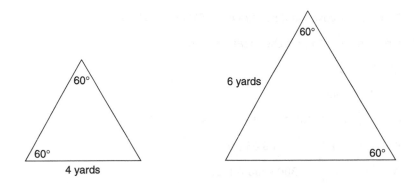

10. A 7-foot upright pole near a vertical tree casts a 6-foot shadow. At the same time of day:

    a. find the height of the tree if its shadow is 36 feet;

    b. find the shadow of the tree if its height is 77 feet.

    (Hint: You should definitely draw diagrams for this problem.)

11. A guy wire attached to a 250-foot tower is fastened to the ground 70 feet from the tower. How long is the guy wire?

12. Find the length of the diagonal of a rectangle the length of which is 10 feet and the width of which is 8 feet.

13. Plot the points (0,2), (−3,2), and (−3,−2), and show that they form the vertices of a right triangle.

14. In softball, the distance from home plate to first base is 60 feet, as is the distance from first base to second base. If the lines joining home plate to first base and first base to second base form a right angle, how far does a catcher standing on home plate have to throw the ball so that it reaches the short-stop standing on second base?

15. A boat is being pulled into a dock with a rope attached at water level. When the boat is 12 feet from the dock, the length of the rope from the boat to the dock is 3 feet longer than twice the height of the dock above water. Find the height of the dock.

1. First we have to find the perimeter: $P = 20 + 23 + 29 = 72$ feet.

    To find the cost, we'll multiply: $C = \$15(72) = \$1,080$.

2. First we have to find the measure of the hypotenuse of the triangle.

    $g^2 = 20^2 + 30^2$

    $g^2 = 400 + 900$

    $g = \sqrt{1,300} = \sqrt{100(13)} = 10\sqrt{13} \approx 36.06$ yards

    $P = 20 + 30 + 10\sqrt{13} \approx 86.06$ yards

    $A = \dfrac{1}{2}\left(\dfrac{30}{1}\right)\left(\dfrac{20}{1}\right) = 300$ square feet

    The cost to fence in the yard: $58(50 + 10\sqrt{13}) = 580(5 + \sqrt{13}) \approx \$4,991.22$

    The cost to cover the lot with sod: $14(300) = \$4,200$

3.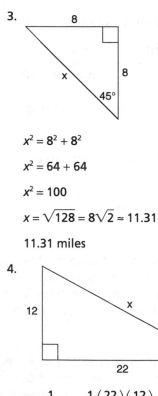

    $x^2 = 8^2 + 8^2$

    $x^2 = 64 + 64$

    $x^2 = 100$

    $x = \sqrt{128} = 8\sqrt{2} \approx 11.31$

    11.31 miles

4.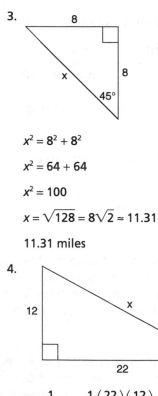

    $A = \dfrac{1}{2}bh = \dfrac{1}{2}\left(\dfrac{22}{1}\right)\left(\dfrac{12}{1}\right) = 132$

    $C = 132(\$10) = \$1,320$

5.

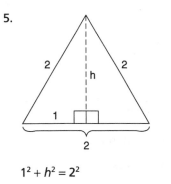

$1^2 + h^2 = 2^2$

$h^2 = 4 - 1$

$h = \sqrt{3}$

$A = \dfrac{1}{2}bh = \dfrac{1}{2}\left(\dfrac{2}{1}\right)\left(\dfrac{\sqrt{3}}{1}\right) = \sqrt{3}$ square inches

$\dfrac{28}{1}\left(\dfrac{\sqrt{3}}{1}\right) = 28\sqrt{3} \approx 48.5$ plugs needed

6.

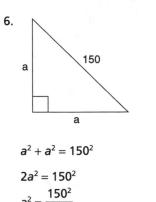

$a^2 + a^2 = 150^2$

$2a^2 = 150^2$

$a^2 = \dfrac{150^2}{2}$

$a = \sqrt{11{,}250} = 75\sqrt{2} \approx 106.07$

106.07 paces to the west,
then 106.07 paces to the north

7. C′D′ = 3(2) = 6 miles

$D'E' = 3\left(2\dfrac{1}{4}\right) = \dfrac{3}{1}\left(\dfrac{9}{4}\right) = \dfrac{27}{4} = 6\dfrac{3}{4}$ inches

C′E′ = 3(2) = 6 inches

$P = 6 + 6\dfrac{3}{4} + 6 = 18\dfrac{3}{4}$ inches

8.

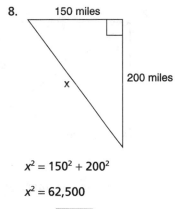

$x^2 = 150^2 + 200^2$

$x^2 = 62{,}500$

$x = \sqrt{62{,}500} = 250$ miles

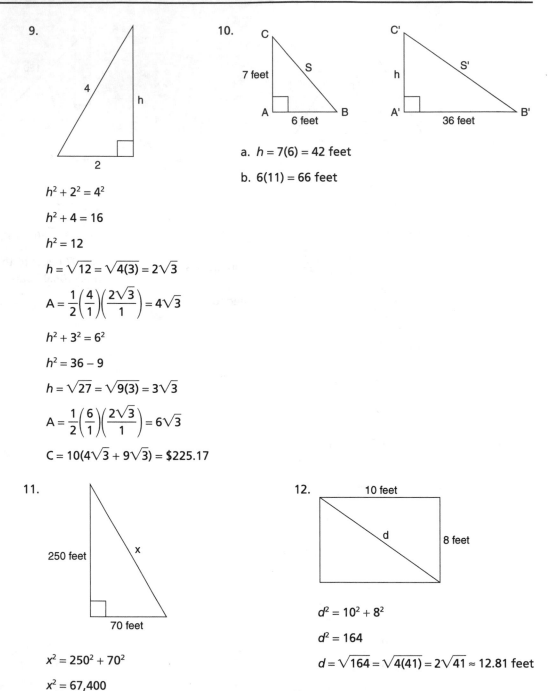

9.

$h^2 + 2^2 = 4^2$

$h^2 + 4 = 16$

$h^2 = 12$

$h = \sqrt{12} = \sqrt{4(3)} = 2\sqrt{3}$

$A = \dfrac{1}{2}\left(\dfrac{4}{1}\right)\left(\dfrac{2\sqrt{3}}{1}\right) = 4\sqrt{3}$

$h^2 + 3^2 = 6^2$

$h^2 = 36 - 9$

$h = \sqrt{27} = \sqrt{9(3)} = 3\sqrt{3}$

$A = \dfrac{1}{2}\left(\dfrac{6}{1}\right)\left(\dfrac{2\sqrt{3}}{1}\right) = 6\sqrt{3}$

$C = 10(4\sqrt{3} + 9\sqrt{3}) = \$225.17$

10.

a. $h = 7(6) = 42$ feet

b. $6(11) = 66$ feet

11.

$x^2 = 250^2 + 70^2$

$x^2 = 67{,}400$

$x = \sqrt{67{,}400} = 10\sqrt{674} \approx 259.62$ feet

12.

$d^2 = 10^2 + 8^2$

$d^2 = 164$

$d = \sqrt{164} = \sqrt{4(41)} = 2\sqrt{41} \approx 12.81$ feet

13.

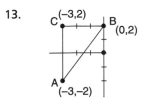

We'll begin by using the distance formula to find the length of each side of the triangle.

$$d = \sqrt{(x_2 - x_1)^2 + (y_2 - y_1)^2}$$

$$d_{CB} = \sqrt{(2 - 2)^2 + (0 - -3)^2} = \sqrt{9} = 3$$

$$d_{CA} = \sqrt{(2 - -2)^2 + (-3 - -3)^2} = \sqrt{4^2} = 4$$

$$d_{BA} = \sqrt{(2 - -2)^2 + (0 - -3)^2} = \sqrt{4^2 + 3^2} = \sqrt{25} = 5$$

Now, using the Pythagorean theorem,

$$(d_{CB})^2 + (d_{CA})^2 = (d_{BA})^2$$

$$3^2 + 4^2 = 5^2$$

$$9 + 16 = 25$$

$$25 = 25$$

14.

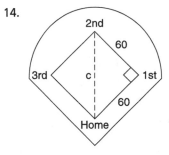

$$c^2 = 60^2 + 60^2$$

$$c^2 = 7{,}200$$

$$c = \sqrt{7{,}200} = \sqrt{3{,}600(2)} = 60\sqrt{2} \approx 84.85 \text{ feet}$$

15.

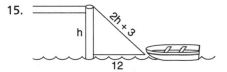

$$h^2 + 12^2 = (2h + 3)^2$$

$$h^2 + 144 = (2h + 3)(2h + 3)$$

$$h^2 + 144 = 4h^2 + 6h + 6h + 9$$

$$0 = 3h^2 + 12h - 135$$

$$0 = 3(h^2 + 4h - 45) = 3(h + 9)(h - 5)$$

$h = -9$, which is not an acceptable answer

$h = 5$ feet

3 Circles

In this chapter we'll study circles and their applications. When you've completed this chapter, you should be able to work with:

- area and circumference
- arc length
- radians and degrees
- the unit circle and basic trigonometry
- writing and graphing equations of circles
- applications involving circles

PRETEST

Now we're going to find out how much you know about circles. If you do really well, you'll be able to skip some of this chapter.

1. If the diameter of a circle is 8 inches, what is its radius?

2. If the radius of a circle is 8 yards, what is its diameter?

3. If the diameter of a circle is 8 inches, what is its circumference?

4. If the radius of a circle is 9 inches, what is its circumference?

5. If the circumference of a circle is 12 yards, what is its diameter?

6. Identify all the chords, arcs, and central angles in the following circle.

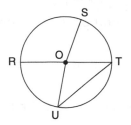

7. Find the area of a circle with a radius of 7 inches.

8. Find the area of a circle with a diameter of 10 cm.

9. Convert the following angles measured in degrees to radian form.

    a. 240°   b. 315°   c. 120°   d. 22°   e. 780°

10. Convert the following angles measured in radians to degree form.

    a. $\dfrac{\pi}{2}$ b. $\dfrac{3\pi}{5}$ c. $\dfrac{\pi}{18}$ d. $\dfrac{7\pi}{10}$ e. $\dfrac{11\pi}{5}$

11. Draw the following angles on a circle.

    a. 150°   b. 450°   c. −65°   d. −450°

12. Draw the following angles on a circle.

    a. $\dfrac{2\pi}{3}$ b. $\dfrac{13\pi}{6}$ c. $-\dfrac{\pi}{2}$ d. $-\dfrac{13\pi}{6}$

13. State two coterminal angles, one positive and one negative, for 32°.

14. State two coterminal angles, one positive and one negative, for −60°.

15. State two coterminal angles, one positive and one negative, for $\dfrac{6\pi}{7}$.

16. State two coterminal angles, one positive and one negative, for $-\dfrac{\pi}{4}$.

17. In the following circle, assuming O is at the center, find:

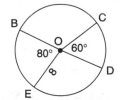

    a. the radius, diameter, circumference, and area of ⊙ O.

    b. the lengths of $\overline{BD}$ and $\overline{OC}$;

    c. the arc length of $\overset{\frown}{ED}$;

    d. the measure of ∠BOC.

18. Find the distance between the points $\left(-\dfrac{1}{2},\dfrac{3}{4}\right)$ and $\left(\dfrac{1}{2},\dfrac{9}{4}\right)$.

19. A circle has a center point at $(4,-5)$ and a point on its circumference at $(1,3)$. Find the radius of the circle.

20. A circle has a center point at $(-5,5)$ and a radius of 3. Find a point on its circumference.

21. Given the circle with center point $(-1,2)$ and radius 4, write the equation of the circle:

    a. in standard form;

    b. in general form.

22. Graph the circle whose equation is $(x-2)^2 + (y+3)^2 = 4$. Label the center point and four points on its circumference.

23. Find the center and radius of the following circle.

    $2x^2 + 2y^2 - 6x + 10y = 1$

    Write the equation in standard form. Graph the circle and label the center point and four points on the circumference of the circle.

24. Find the equation of the circle in which a diameter has end points $P_1 = (-2,1)$ and $P_2 = (2,3)$. Write the equation in:

    a. standard form;

    b. general form.

25. A circular path 5 feet wide is built around a circular garden with a radius of 15 feet. Find the area of the circular path.

26. Find the cost to fence in a circular garden with a diameter of 10 feet if the fencing costs $4.50 per foot.

ANSWERS

1. $r = \dfrac{1}{2}d = \dfrac{1}{2}(8) = 4$ inches

2. $d = 2r = 2(8) = 16$ yards

3. $C = d\pi = 8\pi \approx 25.13$ inches

4. $C = 2\pi r = 2\pi(9) = 18\pi \approx 56.55$ inches

5. $C = d\pi$

   $12 = d\pi$

   $d = \dfrac{12}{\pi} \approx 3.82$ yards

6. Chords: $\overline{UT}, \overline{RT}$

   Arcs: $\overset{\frown}{ST}, \overset{\frown}{TU}, \overset{\frown}{UR}, \overset{\frown}{RS}$

   Central angles: $\angle SOT, \angle TOU, \angle UOR, \angle ROS$

7. $A = \pi(7)^2 = 49\pi \approx 153.94$ square inches

8. $r = \frac{1}{2}d = \frac{1}{2}(10) = 5$

   $A = \pi(5)^2 = 25\pi \approx 78.54 \ cm^2$

9. a. $240\left(\dfrac{\pi}{180}\right) = \dfrac{4\pi}{3}$ b. $315\left(\dfrac{\pi}{180}\right) = \dfrac{7\pi}{4}$ c. $120\left(\dfrac{\pi}{180}\right) = \dfrac{2\pi}{3}$

   d. $22\left(\dfrac{\pi}{180}\right) = \dfrac{11\pi}{90}$ e. $780\left(\dfrac{\pi}{180}\right) = \dfrac{13\pi}{3}$

10. a. $\dfrac{\pi}{2}\left(\dfrac{180}{\pi}\right) = 90°$ b. $\dfrac{3\pi}{5}\left(\dfrac{180}{\pi}\right) = 108°$ c. $\dfrac{\pi}{18}\left(\dfrac{180}{\pi}\right) = 10°$

    d. $\dfrac{7\pi}{10}\left(\dfrac{180}{\pi}\right) = 126°$ e. $\dfrac{11\pi}{5}\left(\dfrac{180}{\pi}\right) = 396°$

11. a.
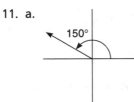

b.
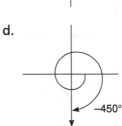

c.
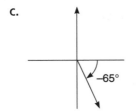

d.
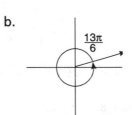

12. a.
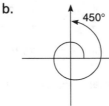

b.
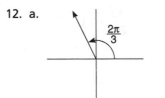

c.

d.
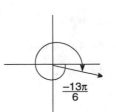

13. Answers may vary, $32 + 360 = 392°$, $32 - 360 = -328°$

    general form: $32 + 360n$, $32 - 360n$, n is an integer.

14. Answers may vary, $-60 + 360 = 300°$, $-60 - 360 = -420°$

    general form: $-60 + 360n$, $-60 - 360n$, n is an integer.

15. Answers may vary, $\dfrac{6\pi}{7} + 2\pi = \dfrac{6\pi}{7} + \dfrac{14\pi}{7} = \dfrac{20\pi}{7}$

    $\dfrac{6\pi}{7} - 2\pi = \dfrac{6\pi}{7} - \dfrac{14\pi}{7} = -\dfrac{8\pi}{7}$

    general form: $\dfrac{6\pi}{7} + 2\pi n$, $\dfrac{6\pi}{7} - 2\pi n$, n is an integer.

16. Answers may vary, $-\dfrac{\pi}{4} + 2\pi = -\dfrac{\pi}{4} + \dfrac{8\pi}{4} = \dfrac{7\pi}{4}$, $-\dfrac{\pi}{4} - 2\pi = -\dfrac{\pi}{4} - \dfrac{8\pi}{4} = -\dfrac{9\pi}{4}$

    general form: $-\dfrac{\pi}{4} + 2\pi n$, $-\dfrac{\pi}{4} - 2\pi n$, n is an integer.

17. a. radius = 8, diameter = 16, $C = 2\pi(8) = 16\pi \approx 50.27$, $A = \pi(8)^2 = 64\pi \approx 201.06$

    b. $\overline{BD} = 16$, $\overline{OC} = 8$

    c. $\angle BOD = 180°$, therefore $\angle EOD = 180° - 80° = 100°$

    $100°\left(\dfrac{\pi}{180}\right) = \dfrac{5\pi}{9}$

    $s = r\theta = 8\left(\dfrac{5\pi}{9}\right) = \dfrac{40\pi}{9} \approx 13.96$

    d. $\angle BOC = 180° - 60° = 120°$

18. $d = \sqrt{\left(\dfrac{1}{2} - -\dfrac{1}{2}\right)^2 + \left(\dfrac{9}{4} - \dfrac{3}{4}\right)^2} = \sqrt{1 + \dfrac{9}{4}} = \sqrt{\dfrac{13}{4}} = \dfrac{\sqrt{13}}{2} \approx 1.8$

19. $r = \sqrt{(1 - 4)^2 + (3 + 5)^2} = \sqrt{9 + 64} = \sqrt{73} \approx 8.5$

20. $(-5,8)$ or $(-5,2)$ or $(-8,5)$ or $(-2,5)$

21. a. $(x + 1)^2 + (y - 2)^2 = 16$

    b. $x^2 + 2x + 1 + y^2 - 4y + 4 = 16$

    $x^2 + y^2 + 2x - 4y - 9 = 0$

22.

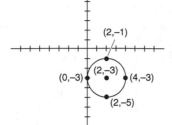

23. $2x^2 + 2y^2 - 6x + 10y = 1$

$x^2 - 3x + y^2 + 5y = \dfrac{1}{2}$

$\left(x^2 - 3x + \dfrac{9}{4}\right) + \left(y^2 + 5y + \dfrac{25}{4}\right) = \dfrac{1}{2} + \dfrac{9}{4} + \dfrac{25}{4}$

$\left(x - \dfrac{3}{2}\right)^2 + \left(y + \dfrac{5}{2}\right)^2 = \dfrac{36}{4}$

Center $= \left(\dfrac{3}{2}, -\dfrac{5}{2}\right)$, radius: 3

Four points: $\left(\dfrac{9}{2}, -\dfrac{5}{2}\right)$, $\left(-\dfrac{3}{2}, -\dfrac{5}{2}\right)$, $\left(\dfrac{3}{2}, \dfrac{1}{2}\right)$, and $\left(\dfrac{3}{2}, -\dfrac{11}{2}\right)$

24. center point $= \left(\dfrac{-2+2}{2}, \dfrac{1+3}{2}\right) = (0,2)$

$r = \sqrt{(2+2)^2 + (3-1)^2} = \sqrt{25} = 5$

a. $x^2 + (y - 2)^2 = 5^2$

b. $x^2 + y^2 - 4y + 4 = 25$

$x^2 + y^2 - 4y - 21 = 0$

25. $A = \pi(20)^2 = 400\pi$, $A = \pi(15)^2 = 225\pi$

area of the circular path $= 400\pi - 225\pi = 175\pi \approx 549.78$

26. $C = 10\pi$, cost $= \$4.50(10\pi) \approx \$141.37$

If you got every problem right, you may go directly to chapter 4. You know what? We're feeling pretty good today, so even if you got one or two wrong answers, we'll still give you permission to skip this chapter.

Now let's get down to some specifics. If you got the right answers for questions 1 and 2, you may skip the "Definition, Radius, and Diameter" section. If you got questions 3 to 6, 9, 10, and 17b, c, and d right, you may skip the "Circumference, Arc Length, Degrees, and Radians" section. If you got questions 7, 8, and 17a correct, you may skip the "Area" section. If you got questions 11 to 16 correct, you may skip the "Angles" section. If you got questions 18 to 24 correct, you may skip the "Graphs and Equations of Circles" section. If you got questions 25 and 26 correct, you may skip the "Applications" section.

Definition, Radius, and Diameter

The circle is a very familiar geometric shape. Indeed, we use the word *circle* in our everyday lives—traffic circle, going around in circles, let's circle around. More formally, *a circle is a planar shape consisting of a closed curve in which each point on the curve is the same distance from the center.*

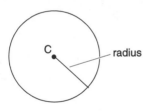

The circle shown above has a straight line called a radius. *The radius of a circle is the distance between the center and any point on the circle.* By definition, all radii (plural of radius) drawn for a given circle have the same length.

In the next circle we've labeled the diameter, which is a *line segment drawn through the center point with its end points on the circumference of the circle.* As you can see, the diameter is twice the radius, d = 2r. The radius is one-half the diameter, $r = \frac{1}{2}d$.

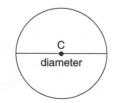

Example 1:

Find the diameter of a circle with a radius of 10 inches.

Solution:

$d = 2r$
$d = 2(10) = 20$ inches.

Example 2:

Find the radius of a circle with a diameter of 8 feet.

Solution:

$$r = \frac{1}{2}d$$

$$r = \frac{1}{2}\left(\frac{8}{1}\right) = 4 \text{ feet}$$

Circumference, Arc Length, Degrees, and Radians

When we look at a circle, what we see is really its circumference. *The circumference of a circle is the distance around it.* It's like the perimeter of a square, rectangle, or triangle, only it's a circle. The intersection between two radii of a circle creates an angle of the circle. A full circle is 360°. If we think of a circle as a pie, an angle of 1° is the size of a pie that has been cut into 360 equal slices. The following illustration shows a circle divided into four equal slices. When the degrees of a circle are measured counterclockwise, the angles are positive. When the degrees are measured clockwise, the angles are negative.

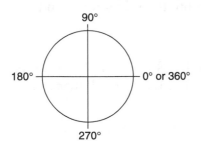

Pi, or π, defines the ratio between the circumference and the diameter of a circle. *π is equivalent to the circumference divided by the diameter of a circle.* The value of

π is $\dfrac{22}{7}$, or approximately 3.14. Every circle has the same ratio. We can use the value of π to find the circumference of a circle.

Circumference of a circle formula:

$C = 2\pi r$ or $C = \pi d$

Example 3:

If the diameter of a circle is 3 inches, what is its circumference?

Solution:

$C = \pi d$, $C = \pi(3) = 9.42$ inches

Example 4:

If the radius of a circle is 1 foot, what is its circumference?

Solution:

$C = 2\pi r$, $C = 2(3.14)(1) = 6.28$ feet

Example 5:

If the circumference of a circle is 16 inches, what is its diameter?

Solution:

If $C = \pi d$, we can solve this equation for d to find an equation for the diameter of a circle.

$C = \pi d$	Divide both sides of the equation by π.
$\dfrac{C}{\pi} = d$	Now we have an equation for the diameter of a circle.
$\dfrac{16}{3.14} = d$	We'll substitute the values for C and π.

$d \approx 5.1$ inches

The symbol we'll <u>use</u> for a <u>circle</u> is $\odot$. *A chord is a line joining any two points on the circumference.* Thus *AB* and *AC* are chords of the following circle.

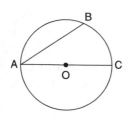

An arc is part of the circumference of a circle. The symbol for arc is $\frown$. Thus $\overset{\frown}{AB}$ refers to arc $\overset{\frown}{AB}$. *An arc of 1° is $\frac{1}{360}$th of a circle.* It's important to distinguish between chords and arcs named with the same letters. In the previous circle, chord $\overline{AB}$ is a straight line joining points A and B, which are both on the circumference of the circle. But arc $\overset{\frown}{AB}$ is actually the curved part of the circumference running from point A to point B.

A diameter is a chord that runs through the center of the circle; it's the longest possible chord. Thus in the last figure, $\overline{AC}$ is the diameter of $\odot$ O.

The diameter divides a circle into two semicircles. *A semicircle is an arc equal to one-half of the circumference of a circle; a semicircle contains* 180°.

Example 6:
Draw a chord, $\overline{CD}$, on $\odot$ M.

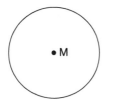

Solution:
Here are two of the many possible chords.

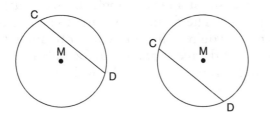

Are you ready for some congruent circles? *Congruent circles are circles having congruent radii.* Thus if OE = O'G', then circle O $\cong$ circle O'.

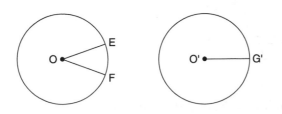

Here's one last term: central angle. *A central angle is an angle formed by two radii.* In the following figure, the angle between radii OB and OC is a central angle. Therefore, the central angle between OB and OC is $\angle O$, which can also be written $\angle BOC$.

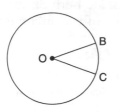

Sometimes angles are not represented in degrees, but in units called *radians*. If a central angle of a circle intercepts an arc equal in length to the radius of the circle, the central angle is defined as 1 radian.

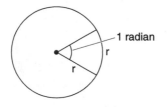

Because the radius can be marked off along the circumference 2π (or about 6.28) times, we see that $2\pi = 360°$, or $\pi = 180°$. If the symbol $°$ isn't written, assume that the angle is represented in radians, not degrees.

The formula to find the measure of a central angle, θ, is also used to find an arc length or the radius of a circle. The formula assumes the angle is written in radian form, not degree form. (If you're using a calculator to apply this formula, make sure it's in radian mode, not degree mode.) The formula is $s = r\theta$, where s is the length of an arc of a circle, r is the radius, and θ is the size of the angle.

Example 7:

Find the measure of $\overset{\frown}{BC}$, if the radius of the circle is 4 and the measure of the central angle is $\dfrac{\pi}{3}$.

Solution:

$$s = r\theta \qquad s = \frac{4}{1}\left(\frac{\pi}{3}\right) = \frac{4\pi}{3} \approx 4.19$$

Example 8:

Find the measure of $\angle BOC$ if $\overset{\frown}{BC} = 2\pi$ and the radius is 6.

Solution:

From the basic formula $s = r\theta$, we can derive a formula to find θ.

$$\theta = \frac{s}{r} = \frac{2\pi}{6} = \frac{\pi}{3}$$

Example 9:

Find the measure of the radius of a circle with a sector that has an arc length of $\dfrac{5\pi}{4}$ and an angle of $\dfrac{\pi}{4}$ units.

Solution:

From the basic formula $s = r\theta$, we can derive a formula to find the radius.

$$r = \frac{s}{\theta} = \frac{\dfrac{5\pi}{4}}{\dfrac{\pi}{4}} = \frac{5\pi}{4}\left(\frac{4}{\pi}\right) = 5$$

When the measure of an angle is given in degree form, you have to convert it to radian form before you can use the formula $s = r\theta$. Don't worry; to convert from degree form to radian form, all you have to do is multiply the angle in degree form by $\dfrac{\pi}{180}$. To convert from radian form to degree form, just multiply the angle in radian form by $\dfrac{180}{\pi}$.

Example 10:

Convert the following commonly used angles to radian form.

a. 0°	**b. 30°**	**c. 45°**	**d. 60°**
e. 90°	**f. 180°**	**g. 270°**	**h. 360°**

Solution:

Multiply each angle by $\dfrac{\pi}{180}$ and reduce.

a. $0°\left(\dfrac{\pi}{180}\right) = 0°$

b. $\dfrac{30°}{1}\left(\dfrac{\pi}{180}\right) = \dfrac{\pi}{6}$

c. $\dfrac{45°}{1}\left(\dfrac{\pi}{180}\right) = \dfrac{\pi}{4}$

d. $\dfrac{60°}{1}\left(\dfrac{\pi}{180}\right) = \dfrac{\pi}{3}$

e. $\dfrac{90°}{1}\left(\dfrac{\pi}{180}\right) = \dfrac{\pi}{2}$

f. $\dfrac{180°}{1}\left(\dfrac{\pi}{180}\right) = \pi$

g. $\dfrac{270°}{1}\left(\dfrac{\pi}{180}\right) = \dfrac{3\pi}{2}$

h. $\dfrac{360°}{1}\left(\dfrac{\pi}{180}\right) = 2\pi$

Example 11:

Convert the following angles from radian form to degree form.

a. $\dfrac{5\pi}{4}$ b. $\dfrac{7\pi}{4}$ c. $\dfrac{6\pi}{5}$ d. $\dfrac{4\pi}{3}$ e. $\dfrac{16\pi}{3}$

Solution:

Multiply each angle by $\dfrac{180}{\pi}$.

a. $\dfrac{5\pi}{4}\left(\dfrac{180}{\pi}\right) = 225°$ b. $\dfrac{7\pi}{4}\left(\dfrac{180}{\pi}\right) = 315°$ c. $\dfrac{6\pi}{5}\left(\dfrac{180}{\pi}\right) = 216°$

d. $\dfrac{4\pi}{3}\left(\dfrac{180}{\pi}\right) = 240°$ e. $\dfrac{16\pi}{3}\left(\dfrac{180}{\pi}\right) = 960°$

Example 12:

Use the following circle to find:

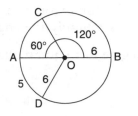

a. the length of $\overline{OC}$,

b. the length of $\overset{\frown}{AB}$

c. the measure of $\angle AOD$,

d. the measure of $\overset{\frown}{AC}$

Solution:

a. $\overline{OC} = 6$

b. To find the length of $\overset{\frown}{AB}$, we have to find the radius of the circle, which is 6. We also have to find the size of the angle in radian form. Since $\overline{AB}$ is the diameter of the circle, we know the size of the angle is π. Now we'll use the formula $s = r\theta$.

$s = 6\pi$

$\overset{\frown}{AB} = 6\pi \approx 18.85$

c. We'll use a variation of the formula $s = r\theta$ to find the measure of θ.

$$\theta = \frac{s}{r} = \frac{5}{6}$$

$$\angle \text{AOD} = \frac{5}{6} \approx .83 \text{ radians, which is} \approx .83\left(\frac{180}{\pi}\right) \approx 48°$$

d. $s = r\theta = 6\left(\dfrac{\pi}{3}\right) = 2\pi$

$$\overset{\frown}{AC} = 2\pi \approx 6.28$$

SELF-TEST 1

1. If the diameter of a circle is 5 inches, what is its circumference?

2. If the radius of a circle is 2 yards, what is its circumference?

3. If the circumference of a circle is 10 inches, what is its diameter?

4. If the circumference of a circle is 8 feet, what is its radius?

For problems 5 to 9, refer to the following circle.

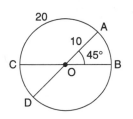

5. Identify all chords, arcs, and central angles.

6. Find the measure of $\overset{\frown}{AB}$.

7. Find the measure of $\angle$AOC.

8. Find the measure of the diameter of the circle.

9. Find the circumference of the circle.

10. Refer to the following circle to answer this problem.

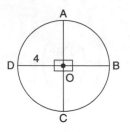

a. Identify all central angles and their values.

b. Find the measure of $\angle DOC$.

c. Find the measure of the diameter of the circle.

d. Find the measure of $\overset{\frown}{AB}$.

e. Find the measure of the circumference of the circle.

ANSWERS

1. $r = \dfrac{1}{2}d = \dfrac{1}{2}\left(\dfrac{5}{1}\right) = \dfrac{5}{2}$ $C = 2\pi r = 2\pi\dfrac{5}{2} = 5\pi \approx 15.71$ inches

2. $C = 2\pi r = 2\pi(2) = 4\pi \approx 12.57$ yards

3. $d = 2r$, $C = 2\pi r$ or $C = d\pi$, $10 = d\pi$, $d = \dfrac{10}{\pi} \approx 3.18$ inches

4. $C = 2\pi r$, $8 = 2\pi r$, $r = \dfrac{8}{2\pi} = \dfrac{4}{\pi} \approx 1.27$ feet

5. chords: $\overline{AD}$, $\overline{BC}$

   arcs: $\overset{\frown}{AB}$, $\overset{\frown}{BD}$, $\overset{\frown}{DC}$, $\overset{\frown}{CA}$, $\overset{\frown}{AD}$, $\overset{\frown}{BC}$

   central angles: $\angle AOB$, $\angle BOD$, $\angle DOC$, $\angle COA$

6. First, convert $45°$ to its radian form, $\dfrac{\pi}{4}$.

   $s = r\theta = \dfrac{10}{1}\left(\dfrac{\pi}{4}\right) = \dfrac{5\pi}{2} \approx 7.85$

7. $\theta = \dfrac{s}{r} = \dfrac{20}{10} = 2$ radians, which is about $\dfrac{2}{1}\left(\dfrac{180}{\pi}\right) = \dfrac{360}{\pi} \approx 115°$

8. $d = 2(10) = 20$

9. $C = 2\pi r = 2\pi(10) = 20\pi \approx 62.83$

10. a. central angles: $\angle AOB$, $\angle BOC$, $\angle COD$, $\angle DOA = 90°$

b. To find the measure of $\angle DOC$, first find the measure of $\overset{\frown}{DC}$. We'll do that by finding the circumference of the circle and dividing it by 4 because this circle is divided into four equal parts. $C = 2\pi r = 2\pi(4) = 8\pi$.

The arc length of $\overset{\frown}{DC}$ is $\dfrac{8\pi}{4} = 2\pi$. $\theta = \dfrac{s}{r} = \dfrac{2\pi}{4} = \dfrac{\pi}{2}$, or 90°. A faster and easier way to find the measure of $\angle DOC$ is to remember that a circle is 360°, which when divided into four equal pieces creates four angles that each must be 90°.

c. $d = 2r = 2(4) = 8$

d. $s = r\theta = \dfrac{4}{1}\left(\dfrac{\pi}{2}\right) = 2\pi$

e. $C = 2\pi r = 2\pi(4) = 8\pi \approx 25.13$

Area

If you happen to know the radius or the diameter of a circle, you can easily find its area. Recall the formula for the area of a circle, $A = \pi r^2$.

It isn't often that mathematical concepts lend themselves to jokes, even really bad ones. So we can't help ourselves here. It seems that when our economy was mired in a terrible depression, every mathematician was thrown out of work. Desperate to find a job, one mathematician applied for a job as a dishwasher. The owner of the restaurant was not only mathematically illiterate, but his grammar was nothing to write home about either. "Tell me a formula," he said to the mathematician. "πr^2 [Pi r squared]," he replied. "I'm sorry," said the restaurant owner, "pie are round; cake are square."

Back to work.

Example 13:
The radius of a circle is 6 inches. Find the area.

Solution:
$A = \pi r^2 = \pi(6)^2 = 36\pi \approx 113.10$ square inches

Example 14:
The diameter of a circle is 8 feet. Find the area.

Solution:
The formula to find the area of a circle uses the radius, not the diameter.

$r = \dfrac{d}{2} = \dfrac{8}{2} = 4$ feet, $A = \pi r^2 = \pi(4)^2 = 16\pi \approx 50.27$ square feet

Example 15:

If the area of a circle is 100, find its radius.

Solution:

$A = \pi r^2$ Substitute 100 for A.

$100 = \pi r^2$ Divide both sides of the equation by π.

$\dfrac{100}{\pi} = r^2$ Take the positive square root of both sides of the equation.

$r = \sqrt{\dfrac{100}{\pi}} \approx 5.64$

And now for something just a little bit different. What if we wanted to find the area of just part of a circle?

For examples 16 and 17, refer to the following circle.

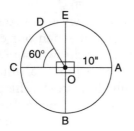

Example 16:

See if you can find the area of section AOB of the circle.

Solution:

We know that a circle contains 360°, and we can see that the area bounded by AOB contains 90°, or one-quarter of the area of the circle. To find that area, we modify our formula:

$A = \dfrac{1}{4}\pi r^2$ Substitute 10 for *r*.

$A = \dfrac{1}{4}\pi (10)^2 \approx 78.54$ square inches

Example 17:

Find the area of section COD of the circle.

Solution:
The area bounded by COD of the circle is $\dfrac{60°}{360°}$, or $\dfrac{1}{6}$th of the circle. To find the area, we modify our formula.

$$A = \frac{1}{6}\pi r^2 \qquad\qquad\qquad \text{Substitute 10 for } r.$$

$$A = \frac{1}{6}\pi(10)^2 \approx 52.36 \text{ square inches}$$

Now we're going to ask you to solve a somewhat complicated problem.

Example 18:

Proportionately, how much smaller is the area of ⊙ M than the area of ⊙ N in the following figure? In other words, what is the ratio of their areas?

Solution:
Let's start by substituting the values for *r* into the formula for the area of a circle.

Area ⊙ M $= \pi r^2 = \pi(5)^2 = 25\pi \approx 78.54$

Area ⊙ N $= \pi r^2 = \pi(10)^2 = 100\pi \approx 314.16$

Can you figure out the ratio of the area of ⊙ M and ⊙ N? It comes to 1:4 $\left(\dfrac{78.54}{314.16} = \dfrac{1}{4}\right)$. Are you surprised? Did you think it would be 1:2 because the ratio of the radii is 1:2? Be careful not to be fooled.

SELF-TEST 2

1. If the radius of a circle is 15 inches, find the circumference.

2. If the diameter of a circle is 10 inches, find the circumference.

3. If the circumference of a circle is 12 inches, find the radius.

4. If the circumference of a circle is 9 feet, find the diameter.

5. If the area of a circle is 9π, find the radius.

6. If the area of a circle is 121π, find the diameter.

Refer to the following circle to answer questions 7 to 9.

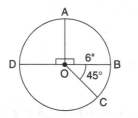

7. Find the area of ⊙ O bounded by AOB.

8. Find the area of ⊙ O bounded by BOC.

9. Find the area of ⊙ O bounded by COD.

10. In the following figure, if the area of ⊙ O is 105 square inches, find:

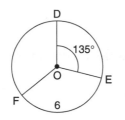

a. the measure of $\overline{OE}$,

b. the measure of $\overset{\frown}{DE}$,

c. the measure of $\angle EOF$. Write your answer in degree form.

11. Refer to the next figure. Proportionately, how much larger is the area of ⊙ M than the area of ⊙ N? In other words, what is the ratio of their areas?

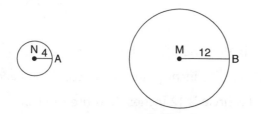

ANSWERS

1. $C = 2\pi r = 2\pi(15) = 30\pi \approx 94.25$ inches   2. $C = 2\pi r = d\pi = 10\pi \approx 31.42$ inches

3. $C = 2\pi r$ Substitute 12 for C.

    $12 = 2\pi r$ Divide both sides of the equation by 2π.

    $r = \dfrac{12}{2\pi} = \dfrac{6}{\pi} \approx 1.91$ inches

4. $C = d\pi$ Substitute 9 for C.

    $9 = d\pi$ Divide both sides of the equation by π.

    $d = \dfrac{9}{\pi} \approx 2.86$ feet

5. $A = \pi r^2$ 6. $A = \pi r^2$

    $9\pi = \pi r^2$ $121\pi = \pi r^2$

    $r^2 = \dfrac{9\pi}{\pi}$ $r^2 = \dfrac{121\pi}{\pi} = 121$

    $r^2 = 9$ $r = \sqrt{121} = 11$

    $r = \sqrt{9} = 3$ $d = 2(11) = 22$

7. $A = \dfrac{1}{4}\pi(6)^2 = 9\pi \approx 28.27$ 8. $\dfrac{45°}{360°} = \dfrac{1}{8}$ $A = \dfrac{1}{8}\pi(6)^2 = \dfrac{1}{8}\pi(36) = \dfrac{9}{2}\pi \approx 14.14$

9. To find the measure of $\angle COD$, we have to notice that $\overline{DB}$ is a diameter of the circle; therefore, $\angle BOD$ is 180° and $\angle COD = 180° - 45° = 135°$. To find the portion of the circle represented by $\angle COD$, divide 135° by 360°.

    $\dfrac{135°}{360°} = \dfrac{3}{8}$ COD is $\dfrac{3}{8}$th of the circle.

    $A = \dfrac{3}{8}\pi(6)^2 = \dfrac{27\pi}{2} \approx 42.41$ inches

10. a. $A = \pi r^2$

    $105 = \pi(OE)^2$

    $\dfrac{105}{\pi} = (OE)^2$

    $OE = \sqrt{\dfrac{105}{\pi}} \approx 5.78$ OE is a radius of the circle.

    b. $135°\left(\dfrac{\pi}{180°}\right) = \dfrac{3\pi}{4}$

    $s = r\theta = (5.78)\left(\dfrac{3\pi}{4}\right) \approx 13.62$

c. $s = r\theta$

$6 = 5.78\theta$

$\theta = \dfrac{6}{5.78} \approx 1.04$ radians

$(1.04 \text{ radians})\left(\dfrac{180}{\pi}\right) \approx 59.59°$

11. $A_N = \pi(4)^2 = 16\pi$ $A_M = \pi(12)^2 = 144\pi$

$\dfrac{16\pi}{144\pi} = \dfrac{1}{9}$

The ratio of the areas of ⊙ M to ⊙ N is 1:9, or one to nine.

Angles

In chapter 2 we talked a little bit about angles. Earlier in this chapter we explained that we can use two forms to denote angles: degrees and radians. We showed you how to convert from radian form to degree form and from degree form to radian form. In this section we'll give you a better explanation of what a radian is and teach you how to draw angles in both degree and radian form. We'll also show you how to find something called coterminal angles, and we'll introduce you to the unit circle.

Let's go back to the idea of a radian as a measurement of the size of an angle. *If a central angle of a circle intercepts an arc equal in length to the radius of the circle, the central angle is defined as one radian.*

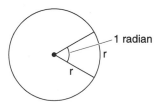

Because the radius can be marked off along the circumference 2π (about 6.28) times, we can say that $2\pi = 360°$, or $\pi = 180°$. If the degree symbol (°) isn't written, it's assumed that the angle is in radians. Now it's time for you to learn how to draw angles in both degree and radian form and how to find terminal angles in both forms.

We know that one complete revolution around a circle is 360°, so a quarter of the way around a circle is 90°, halfway around is 180°, and three-quarters around is 270°. An angle is measured from its initial side to its terminal side. The following circle is called a unit circle centered at the origin (0,0). *A unit circle is a circle with a radius of 1 unit.* Its area is π and its circumference is 2π. The unit circle shown is divided into four equal sectors. Each sector is called a *quadrant*. The quadrants are labeled I, II, III, and IV in a counterclockwise direction.

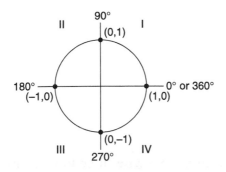

To graph an angle, keep in mind that positive angles move in a counterclockwise direction and negative angles move in a clockwise direction. The angles begin at their initial side, which is the x-axis, and end at what's called their terminal side. We'll use θ to represent an angle without a specified measure. Any angle between 0° and 90° is located in quadrant I, between 90° and 180° in quadrant II, between 180° and 270° in quadrant III, and between 270° and 360° in quadrant IV. The following illustrations show the graphs of angles 60°, 280°, −120°, and −450°.

60° is positive, so it's measured in the counterclockwise direction and between 0° and 90°, which is quadrant I.

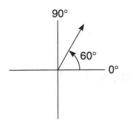

280° is positive, so it's measured in the counterclockwise direction and between 270° and 360°, which is in quadrant IV.

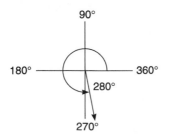

−120° is negative, so it's measured in a clockwise direction and between −90° and −180°, which is quadrant III.

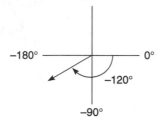

−450° is negative, so it's measured in a clockwise direction. It's more than one complete revolution around the circle. One clockwise revolution is −360°. In order to graph −450°, we have to go −360° and an additional −90°. −450° is not in a quadrant. It's on the *y*-axis.

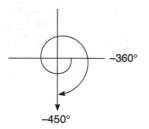

Example 19:

Draw the following angles on one graph: 90°, 450°, and −270°

Solution:

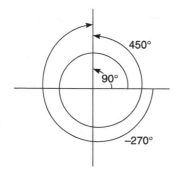

Example 20:

Draw the following angles on one graph: 60°, –300°, –660°

Solution:

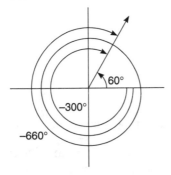

Did you notice that in examples 19 and 20, the angles you graphed have the same terminal side? Angles that have the same terminal side are called *coterminal angles.* Coterminal angles differ by 360° or a multiple of 360° because one complete revolution around a circle is 360°. To find a coterminal angle for any angle, just add or subtract 360°, or any multiple of 360°, to the angle. A more formal way to state this is $\theta \pm 360°n$, where *n* is an integer, is a coterminal angle of θ.

We can find a coterminal angle by adding or subtracting 360°, 720°, 1,080°, and so on.

Example 21:

State three positive coterminal angles for 80°. State three negative coterminal angles for –155°.

Solution:

To find three positive coterminal angles for 80°, all we have to do is add 360° or any multiple of 360° to 80°. Answers may vary, but they all have to be of the form $80° + 360°n$, where *n* is a positive integer. Some acceptable coterminal angles are: $80° + 360° = 440°$, $80° + 360°(2) = 80° + 720° = 800°$, $80° + 360°(3) = 80° + 1,080° = 1,160°$.

To find three negative coterminal angles for –155°, all we have to do is add –360° or any multiple of –360° to –155°. Answers may vary, but they all have to be of the form $-155° - 360°n$, where *n* is a positive integer. Some acceptable coterminal angles are: $-155° - 360° = -515°$, $-155° - 720° = -875°$, $-155° - 1,080° = -1,235°$.

So far all the angles we've worked with have been in degree form. The following illustration shows the unit circle with its angles labeled in radian form.

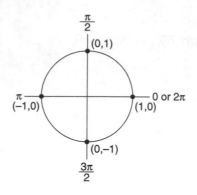

Angles between 0 and $\dfrac{\pi}{2}$ are located in quadrant I, between $\dfrac{\pi}{2}$ and π in quadrant II, between π and $\dfrac{3\pi}{2}$ in quadrant III, and between $\dfrac{3\pi}{2}$ and 2π in quadrant IV. Let's graph some angles in radian form.

Example 22:
Draw the following angles on one graph: $\dfrac{\pi}{2}, \dfrac{8\pi}{3}, -\dfrac{4\pi}{3}, -\dfrac{9\pi}{2}$

Solution:

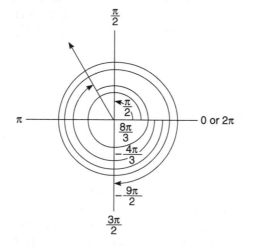

Coterminal angles differ by 2π units, or a multiple of 2π, because one complete revolution around a circle is 2π units. To find a coterminal angle for any angle, just add or subtract 2π or any multiple of 2π to the angle. The following is a more formal way to state this:

$\theta \pm 2\pi n$, where n is an integer, is a coterminal angle for angle θ.

We can find a coterminal angle by adding or subtracting 2π, 4π, 6π, and so on.

Example 23:

State three positive and three negative coterminal angles for each of the following angles: $\dfrac{\pi}{8}, -\dfrac{6\pi}{5}$

Solution:

To find three positive coterminal angles for $\dfrac{\pi}{8}$, all we have to do is add 2π or any multiple of 2π to $\dfrac{\pi}{8}$. Answers may vary, but they all have to be of the form $\dfrac{\pi}{8} + 2\pi n$, where n is a positive integer. Some acceptable positive coterminal angles are: $\dfrac{\pi}{8} + 2\pi = \dfrac{\pi}{8} + \dfrac{16\pi}{8} = \dfrac{17\pi}{8}$, $\dfrac{\pi}{8} + 4\pi = \dfrac{\pi}{8} + \dfrac{32\pi}{8} = \dfrac{33\pi}{8}$, $\dfrac{\pi}{8} + 6\pi = \dfrac{\pi}{8} + \dfrac{48\pi}{8} = \dfrac{49\pi}{8}$

To find three negative coterminal angles for $\dfrac{\pi}{8}$, all we have to do is add -2π or any multiple of -2π to $\dfrac{\pi}{8}$. Answers may vary, but they all have to be of the form $\dfrac{\pi}{8} - 2\pi n$, where n is a positive integer. Some acceptable negative coterminal angles are: $\dfrac{\pi}{8} - 2\pi = \dfrac{\pi}{8} - \dfrac{16\pi}{8} = -\dfrac{15\pi}{8}$, $\dfrac{\pi}{8} - 4\pi = \dfrac{\pi}{8} - \dfrac{32\pi}{8} = -\dfrac{31\pi}{8}$, $\dfrac{\pi}{8} - 6\pi = \dfrac{\pi}{8} - \dfrac{48\pi}{8} = -\dfrac{47\pi}{8}$

SELF-TEST 3

1. Graph the following angles and state the quadrant they're in. Are any of these angles coterminal?

   a. 25°   b. 130°   c. 475°   d. –230°   e. –55°   f. –415°

2. Graph the following angles and state the quadrant they're in. Are any of these angles coterminal?

   a. $\dfrac{3\pi}{4}$ b. $\dfrac{2\pi}{3}$ c. $\dfrac{8\pi}{3}$ d. $-\dfrac{4\pi}{5}$ e. $-\dfrac{7\pi}{6}$ f. $-\dfrac{9\pi}{2}$

3. State one positive and one negative coterminal angle for each of the following angles.

   a. 60°   b. 320°   c. 250°   d. –550°   e. –45°   f. –180°

4. State one positive and one negative coterminal angle for each of the following angles.

   a. $\dfrac{3\pi}{10}$ b. $\dfrac{7\pi}{2}$ c. $\dfrac{11\pi}{3}$ d. $-\dfrac{\pi}{4}$ e. $-\dfrac{7\pi}{2}$ f. $-\dfrac{5\pi}{3}$

5. Which of the following angles is a coterminal angle of 30°?

   a. 180°   b. 390°   c. −330°   d. 330°   e. 1,110°

6. Which of the following angles is a coterminal angle of $\dfrac{\pi}{6}$?

   a. $\dfrac{2\pi}{3}$ b. $\dfrac{13\pi}{6}$ c. $-\dfrac{7\pi}{6}$ d. $-\dfrac{13\pi}{6}$ e. $-\dfrac{11\pi}{6}$

ANSWERS

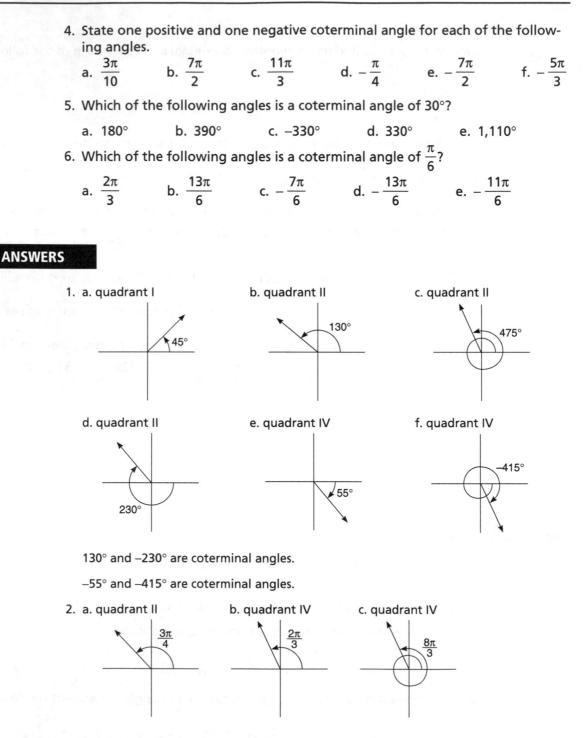

1. a. quadrant I   b. quadrant II   c. quadrant II

   d. quadrant II   e. quadrant IV   f. quadrant IV

   130° and −230° are coterminal angles.

   −55° and −415° are coterminal angles.

2. a. quadrant II   b. quadrant IV   c. quadrant IV

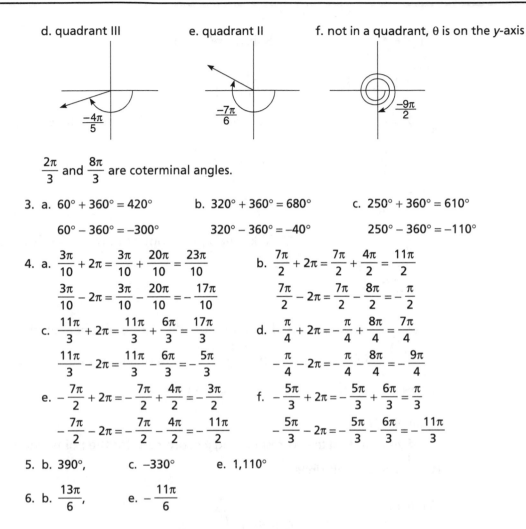

d. quadrant III   e. quadrant II   f. not in a quadrant, θ is on the y-axis

$-\dfrac{4\pi}{5}$ $-\dfrac{7\pi}{6}$ $-\dfrac{9\pi}{2}$

$\dfrac{2\pi}{3}$ and $\dfrac{8\pi}{3}$ are coterminal angles.

3. a. $60° + 360° = 420°$ b. $320° + 360° = 680°$ c. $250° + 360° = 610°$

   $60° - 360° = -300°$ $320° - 360° = -40°$ $250° - 360° = -110°$

4. a. $\dfrac{3\pi}{10} + 2\pi = \dfrac{3\pi}{10} + \dfrac{20\pi}{10} = \dfrac{23\pi}{10}$ b. $\dfrac{7\pi}{2} + 2\pi = \dfrac{7\pi}{2} + \dfrac{4\pi}{2} = \dfrac{11\pi}{2}$

   $\dfrac{3\pi}{10} - 2\pi = \dfrac{3\pi}{10} - \dfrac{20\pi}{10} = -\dfrac{17\pi}{10}$ $\dfrac{7\pi}{2} - 2\pi = \dfrac{7\pi}{2} - \dfrac{8\pi}{2} = -\dfrac{\pi}{2}$

   c. $\dfrac{11\pi}{3} + 2\pi = \dfrac{11\pi}{3} + \dfrac{6\pi}{3} = \dfrac{17\pi}{3}$ d. $-\dfrac{\pi}{4} + 2\pi = -\dfrac{\pi}{4} + \dfrac{8\pi}{4} = \dfrac{7\pi}{4}$

   $\dfrac{11\pi}{3} - 2\pi = \dfrac{11\pi}{3} - \dfrac{6\pi}{3} = \dfrac{5\pi}{3}$ $-\dfrac{\pi}{4} - 2\pi = -\dfrac{\pi}{4} - \dfrac{8\pi}{4} = -\dfrac{9\pi}{4}$

   e. $-\dfrac{7\pi}{2} + 2\pi = -\dfrac{7\pi}{2} + \dfrac{4\pi}{2} = -\dfrac{3\pi}{2}$ f. $-\dfrac{5\pi}{3} + 2\pi = -\dfrac{5\pi}{3} + \dfrac{6\pi}{3} = \dfrac{\pi}{3}$

   $-\dfrac{7\pi}{2} - 2\pi = -\dfrac{7\pi}{2} - \dfrac{4\pi}{2} = -\dfrac{11\pi}{2}$ $-\dfrac{5\pi}{3} - 2\pi = -\dfrac{5\pi}{3} - \dfrac{6\pi}{3} = -\dfrac{11\pi}{3}$

5. b. $390°$,   c. $-330°$ e. $1{,}110°$

6. b. $\dfrac{13\pi}{6}$,   e. $-\dfrac{11\pi}{6}$

Now that you know a little more about the unit circle and angles, we'd like to use your knowledge of the unit circle to give you a very brief introduction to trigonometry. This introduction only scratches the surface. If you would like to learn more about trigonometry, we suggest you review our book *Precalculus: A Self-Teaching Guide*, which is also published by Wiley.

For every point on the circumference of the unit circle, there is an x and y coordinate, (x, y). Luckily, a fairly straightforward equation governs these coordinates: $x^2 + y^2 = 1$. (Does this equation look familiar?) Suppose we want to find the y-coordinate of the point on the circumference of the unit circle where x equals $\dfrac{1}{2}$; to find the corresponding y-coordinate, all we have to do is substitute $\dfrac{1}{2}$ for x into the equation of the circle, $x^2 + y^2 = 1$, and solve for y.

$x^2 + y^2 = 1$ Substitute $\dfrac{1}{2}$ for x.

$\left(\dfrac{1}{2}\right)^2 + y^2 = 1$ $\left(\dfrac{1}{2}\right)^2 = \dfrac{1}{4}$

$\dfrac{1}{4} + y^2 = 1$ Subtract $\dfrac{1}{4}$ from both sides of the equation.

$y^2 = 1 - \dfrac{1}{4}$ Write 1 as $\dfrac{4}{4}$.

$y^2 = \dfrac{4}{4} - \dfrac{1}{4}$ Subtract.

$y^2 = \dfrac{3}{4}$ Take the square root of both sides of the equation.

$y = \sqrt{\dfrac{3}{4}} = \dfrac{\sqrt{3}}{2}$

Now we know one of the points on the circumference of the unit circle is $\left(\dfrac{1}{2}, \dfrac{\sqrt{3}}{2}\right)$. We know the point $\left(\dfrac{1}{2}, \dfrac{\sqrt{3}}{2}\right)$ is in quadrant I because the x-coordinate is positive and the y-coordinate is also positive.

Example 24:

See if you can find the corresponding y-coordinate for the point whose x-coordinate is $-\dfrac{1}{2}$ on the unit circle.

Solution:

$x^2 + y^2 = 1$ Substitute $\left(-\dfrac{1}{2}\right)$ for x.

$\left(-\dfrac{1}{2}\right)^2 + y^2 = 1$ $\left(-\dfrac{1}{2}\right)^2 = \dfrac{1}{4}$

$\dfrac{1}{4} + y^2 = 1$ Subtract $\dfrac{1}{4}$ from both sides of the equation.

$y^2 = 1 - \dfrac{1}{4}$ Write 1 as $\dfrac{4}{4}$.

$y^2 = \dfrac{4}{4} - \dfrac{1}{4}$ Subtract.

$y = \sqrt{\dfrac{3}{4}} = \dfrac{\sqrt{3}}{2}$

Now we know another point on the circumference of the unit circle is $\left(-\dfrac{1}{2}, \dfrac{\sqrt{3}}{2}\right)$. Did you notice that the corresponding y-coordinates of $x = \dfrac{1}{2}$ and $x = -\dfrac{1}{2}$ are the same, $\dfrac{\sqrt{3}}{2}$? We know the point $\left(-\dfrac{1}{2}, \dfrac{\sqrt{3}}{2}\right)$ is in quadrant II

because the x-coordinate is negative and the y-coordinate is positive.

In the following illustration, we have drawn a triangle inside the unit circle.

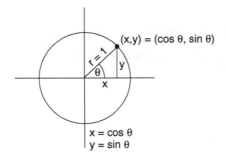

The horizontal distance of the triangle along the x-axis is labeled x. The vertical distance of the triangle to the point labeled (x,y) is labeled y. The angle formed with the origin is labeled θ. Because this is the unit circle, the radius r equals 1. The radius is the hypotenuse of the right triangle. The x-coordinate of the point (x, y) on the unit circle is called the cosine of θ, which we will write as $\cos\theta$. The $\cos\theta = \dfrac{x}{r}$. The y-coordinate of the point (x, y) on the unit circle is called the sine of θ, which we will write as $\sin\theta$. The $\sin\theta = \dfrac{y}{r}$. Cos$\theta$ and $\sin\theta$ are two of the six basic trigonometric functions of an angle, θ, of a right triangle. Another basic trig function is called thetangent of θ, which we will write as $\tan\theta$. The $\tan\theta = \dfrac{\sin\theta}{\cos\theta} = \dfrac{\dfrac{y}{r}}{\dfrac{x}{r}} = \dfrac{y}{r} \cdot \dfrac{r}{x} = \dfrac{y}{x}$.

The following circle shows the signs of the trig functions $\sin\theta$, $\cos\theta$, and $\tan\theta$ in each of the four quadrants. Remember, $y = \sin\theta$, $x = \cos\theta$, and $\tan\theta = \dfrac{y}{x}$.

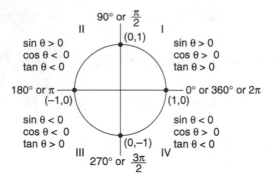

Example 25:

We know the sinθ is the *y*-coordinate of the point (*x*,*y*) on the unit circle, the cosθ is the *x*-coordinate, and the tanθ = $\frac{y}{x}$. Use this knowledge to answer the following questions. In what quadrant(s) is

    a. the sinθ positive?

    b. the cosθ and the sinθ both negative?

    c. the tanθ positive?

    d. the tanθ positive and the sinθ negative?

Solution:

    a. *y* is positive in quadrants I and II, so the sinθ is positive in quadrants I and II.

    b. *x* and *y* are both negative in quadrant III.

    c. In order for the tanθ to be positive, either both *x* and *y* have to be positive, which occurs in quadrant I, or both *x* and *y* have to be negative, which occurs in quadrant III.

    d. The tanθ is positive in quadrants I and III, the sinθ is negative in quadrants III and IV. The tanθ is positive and the sinθ is negative in quadrant III.

Example 26:

Given a point on the unit circle with an *x*-coordinate of $-\frac{\sqrt{3}}{2}$, located in quadrant III, find the sinθ, cosθ, and tanθ.

Solution:

We know $r = 1$ because we're working with the unit circle. We also know $x = -\dfrac{\sqrt{3}}{2}$ because it's given. Now we'll find its corresponding y value by using the equation $x^2 + y^2 = 1$.

$x^2 + y^2 = 1$ Substitute $-\dfrac{\sqrt{3}}{2}$ for x.

$\left(-\dfrac{\sqrt{3}}{2}\right)^2 + y^2 = 1$ Remember $(\sqrt{3})^2 = 3$.

$\dfrac{3}{4} + y^2 = 1$ Subtract $\dfrac{3}{4}$ from both sides of the equation.

$y^2 = 1 - \dfrac{3}{4}$ Write 1 as $\dfrac{4}{4}$.

$y^2 = \dfrac{4}{4} - \dfrac{3}{4}$ Subtract.

$y^2 = \dfrac{1}{4}$ Take the square root of both sides of the equation.

$y = -\sqrt{\dfrac{1}{4}} = -\dfrac{1}{2}$ We had to take the negative square root of $\dfrac{1}{4}$ because the y values are negative in quadrant III.

A point on the circumference of the circle is $\left(-\dfrac{\sqrt{3}}{2}, -\dfrac{1}{2}\right)$. Now that we know the values for x and y, we know the $\sin\theta = \dfrac{y}{r} = -\dfrac{1}{2}$, the $\cos\theta = \dfrac{x}{r} = -\dfrac{\sqrt{3}}{2}$, and the

$\tan\theta = \dfrac{y}{x} = \dfrac{-\dfrac{1}{2}}{-\dfrac{\sqrt{3}}{2}} = -\dfrac{1}{2}\left(-\dfrac{2}{\sqrt{3}}\right) = \dfrac{1}{\sqrt{3}} = \dfrac{\sqrt{3}}{3}.$

SELF-TEST 5

1. In which quadrant(s) is:
   a. the $\sin\theta$ positive?
   b. the $\cos\theta$ negative?
   c. the $\tan\theta$ negative?
   d. the $\tan\theta$ positive and the $\cos\theta$ negative?
   e. the $\sin\theta$ negative, $\cos\theta$ positive, and $\tan\theta$ negative?

2. Given a point on the circumference of the unit circle in quadrant II, find the corresponding y-coordinate if the point's x-coordinate is $-\dfrac{1}{2}$.

3. Given a point on the circumference of the unit circle in quadrant IV, find the corresponding x-coordinate if the point's y-coordinate is $-\dfrac{\sqrt{2}}{2}$.

4. Given a point on the circumference of the unit circle in quadrant I whose x-coordinate is $\dfrac{\sqrt{3}}{2}$, find the $\sin\theta$, $\cos\theta$, and $\tan\theta$.

5. Given a point on the circumference of the unit circle in quadrant II whose y-coordinate is $\dfrac{\sqrt{2}}{2}$, find the $\sin\theta$, $\cos\theta$, and $\tan\theta$.

6. Given the point $(-1,0)$ on the circumference of the unit circle, find the $\sin\theta$, $\cos\theta$, and $\tan\theta$. Where is this point located?

7. Fill in the missing parts of the following table. $0° \leq \theta \leq 360°$

θ	(x,y)	sin θ	cos θ	tan θ
0°	(1,0)	0	1	
90°				
180°				
	(,−1)			
		0	1	

1. a. The $\sin\theta$ is the y-coordinate of the points on the unit circle. The y-coordinates are positive in quadrants I and II.

   b. The $\cos\theta$ is the x-coordinate of the points on the unit circle. The x-coordinates are negative in quadrants II and III.

   c. The $\tan\theta$ is the y-coordinate divided by the x-coordinate. This quotient is negative in quadrants II and IV.

   d. The $\tan\theta$ is positive in quadrants I and III. The $\cos\theta$ is negative in quadrants II and III. The answer is quadrant III.

   e. The $\sin\theta$ is negative in quadrants III and IV. The $\cos\theta$ is positive in quadrants I and IV. The $\tan\theta$ is negative in quadrants II and IV. The answer is quadrant IV.

2. The y-coordinate is positive in quadrant II.

$$x^2 + y^2 = 1$$

$$\left(-\frac{1}{2}\right)^2 + y^2 = 1$$

$$y^2 = 1 - \frac{1}{4}$$

$$y = \sqrt{\frac{3}{4}} = \frac{\sqrt{3}}{2}$$

3. The x-coordinate is positive in quadrant IV.

$$x^2 + y^2 = 1$$

$$x^2 + \left(-\frac{\sqrt{2}}{2}\right)^2 = 1$$

$$x^2 + \frac{2}{4} = 1$$

$$x^2 = 1 - \frac{1}{2}$$

$$x^2 = \frac{1}{2}$$

$$x = \sqrt{\frac{1}{2}} = \frac{\sqrt{2}}{2}$$

4. In quadrant I, the x-coordinate and y-coordinate are both positive.

First we have to find the corresponding y-coordinate to the point whose x-coordinate is $\frac{\sqrt{3}}{2}$.

$$x^2 + y^2 = 1$$

$$\left(\frac{\sqrt{3}}{2}\right)^2 + y^2 = 1$$

$$\frac{3}{4} + y^2 = 1$$

$$y^2 = 1 - \frac{3}{4}$$

$$y^2 = \frac{1}{4}$$

$$y = \sqrt{\frac{1}{4}} = \frac{1}{2}$$

$$\sin\theta = \frac{y}{r} = \frac{\frac{1}{2}}{1} = \frac{1}{2} \qquad \cos\theta = \frac{x}{r} = \frac{\frac{\sqrt{3}}{2}}{1} = \frac{\sqrt{3}}{2} \qquad \tan\theta = \frac{y}{x} = \frac{\frac{1}{2}}{\frac{\sqrt{3}}{2}} = \frac{1}{\sqrt{3}} = \frac{\sqrt{3}}{3}$$

5. In quadrant II the *x*-coordinate is negative and the *y*-coordinate is positive. First we have to find the corresponding *x*-coordinate to the point whose *y*-coordinate is $\frac{\sqrt{2}}{2}$.

$$x^2 + y^2 = 1$$

$$x^2 + \left(\frac{\sqrt{2}}{2}\right)^2 = 1$$

$$x^2 + \frac{2}{4} = 1$$

$$x^2 = 1 - \frac{1}{2}$$

$$x^2 = \frac{1}{2}$$

$$x = -\sqrt{\frac{1}{2}} = -\frac{\sqrt{2}}{2}$$

$$\sin\theta = \frac{\sqrt{2}}{2}, \cos\theta = -\frac{\sqrt{2}}{2}, \tan\theta = \frac{\frac{\sqrt{2}}{2}}{-\frac{\sqrt{2}}{2}} = -1$$

6. The *x*-coordinate is −1, the *y*-coordinate is 0. The $\sin\theta = 0$, $\cos\theta = -1$, $\tan\theta = \frac{0}{-1} = 0$.

7.

θ	(x, y)	sin θ	cos θ	tan θ
0°	(1,0)	0	1	0
90°	(0,1)	1	0	Undefined
180°	(−1,0)	0	−1	0
270°	(0,−1)	−1	0	Undefined
360°	(1,0)	0	1	0

Graphs and Equations of Circles

In this section we'll learn how to graph a circle given its equation and how to write the equation of a circle given its graph. So what are we waiting for? Let's get started!

The following illustration shows a circle with center point $C(h,k)$, radius r, and a point on the circumference of the circle $P(x,y)$. $P(x,y)$ represents any point on the circumference of the circle. The distance from the center of the circle $C(h,k)$ to any point $P(x,y)$ on the circumference of the circle remains constant.

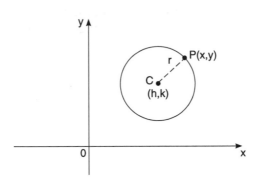

To derive the equation of a circle, we use a variation of the distance formula:

$\sqrt{(x-h)^2 + (y-k)^2} = r$ r is the radius of the circle.

$(\sqrt{(x-h)^2 + (y-k)^2})^2 = r^2$ Square both sides of the equation.

$(x-h)^2 + (y-k)^2 = r^2$ Now we have the formula for the equation of a circle.

The formula for the equation of a circle in standard form is:

$(x-h)^2 + (y-k)^2 = r^2$ where $C(h,k)$ is the center point of the circle and r is the radius of the circle.

Example 27:

See if you can graph the following circle. Label four points on its circumference.

$(x-2)^2 + (y+1)^2 = 9$

Solution:

The center of the circle is $C(2,-1)$ and $r^2 = 9$, so the radius is $\sqrt{9} = 3$. It's easy to find four points on the circumference of the circle by moving three units to the right, left, above, and below the center point, (2,–1). Moving to the right or left of the center point is a horizontal change and therefore changes the *x*-coordinate of (2,–1). Moving up or down from the center point is a vertical change and therefore changes the *y*-coordinate. If we move three units to the right of (2,–1), the *x*-coordinate increases by 3, giving us the point (5,–1), which is on the circumference of the circle. If we move three units to the left of (2,–1), the *x*-coordinate decreases by 3, giving us the point (–1,–1) which is also on the circumference of the circle. If we move three units above (2,–1), the *y*-coordinate increases by 3 units, giving us the point (2,2), which is also on the circumference of the circle. If we move three units below (2,–1), the *y*-coordinate decreases by 3, giving us the point (2,–4), which is also on the circumference of the circle. It's possible to find other points on the circumference of the circle, but these are the easiest ones to find.

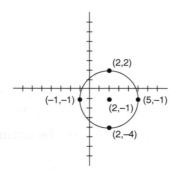

Suppose we gave you the reverse of this problem. If we gave you the graph and asked you to give us the equation of the circle, could you do it?

Example 28:
Write the equation of the previous graph.

Solution:
To write the equation of the circle, we have to use the formula:

$(x – h)^2 + (y – k)^2 = r^2$ Substitute 2 for *h*, –1 for *k*, and 3 for *r*.

$(x – 2)^2 + (y – –1)^2 = 3^2$

$(x – 2)^2 + (y + 1)^2 = 9$

Example 29:

Write the equation of the following graph.

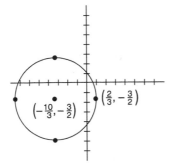

Solution:

To write the equation of the graph, we have to use the formula $(x - h)^2 + (y - k)^2 = r^2$, where (h,k) is the center point and r is the radius of the circle. The center point $\left(-\dfrac{10}{3}, -\dfrac{3}{2}\right)$ of the graph is given, so all we have to find is the length of the radius of the circle. Once we know the length of the radius, we can substitute these values into the standard formula of the equation of a circle.

The center of the graph is $\left(-\dfrac{10}{3}, -\dfrac{3}{2}\right)$, so $h = -\dfrac{10}{3}$ and $k = -\dfrac{3}{2}$. To find the length of the radius, we'll use the distance formula: $d = \sqrt{(x_2 - x_1)^2 + (y_2 - y_1)^2}$. Then substitute the coordinates of the center point $(-\dfrac{10}{3}, -\dfrac{3}{2})$ and any one of the points on the circumference of the circle into the distance formula. We'll arbitrarily choose the point $(\dfrac{2}{3}, -\dfrac{3}{2})$, which is on the circumference of the circle.

$d = \sqrt{(x_2 - x_1)^2 + (y_2 - y_1)^2}$ Substitute the values for the variables.

$d = \sqrt{\left(\dfrac{2}{3} - -\dfrac{10}{3}\right)^2 + \left(-\dfrac{3}{2} - -\dfrac{3}{2}\right)^2}$ Change the double negatives to positives.

$d = \sqrt{\left(\dfrac{2}{3} + -\dfrac{10}{3}\right)^2 + \left(-\dfrac{3}{2} + \dfrac{3}{2}\right)^2}$ Add the fractions.

$d = \sqrt{\left(\dfrac{12}{3}\right)^2 + 0}$ $\dfrac{12}{3} = 4$

$d = \sqrt{(4)^2} = \sqrt{16} = 4$ The radius of the circle is 4.

Now that we know the center point $\left(-\dfrac{10}{3}, -\dfrac{3}{2}\right)$ and the radius, 4, of this circle, we can substitute these values into the formula for the equation of a circle.

$(x - h)^2 + (y - k)^2 = r^2$

$\left(x - -\dfrac{10}{3}\right)^2 + \left(y - -\dfrac{3}{2}\right)^2 = 4^2$ Change the double negatives to positives.

$\left(x + \dfrac{10}{3}\right)^2 + \left(y + \dfrac{3}{2}\right)^2 = 16$ This is the equation of the circle.

Example 30:

Write the equation of the following circle.

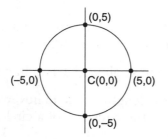

Solution:

The radius of the circle is 5. The center point is (0,0). Now we have to substitute 0 for h, 0 for k, and 5 for r in the standard equation of the circle.

$(x - h)^2 + (y - k)^2 = r^2$

$(x - 0)^2 + (y - 0)^2 = 5^2$

$x^2 + y^2 = 25$

$x^2 + y^2 = r^2$ This is the standard equation of a circle centered at the origin.

SELF-TEST 5

1. Find the radius of the circle with a center point of $(-1,3)$ and a point on its circumference of $(2,1)$.

2. State the equation of the following circle.

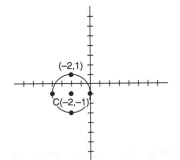

3. Graph the following circle. Label four points on its circumference. This is a special circle you should recognize.

   $x^2 + y^2 = 1$

4. Graph the following circle. Label four points on its circumference.

   $(x - 4)^2 + (y + 5)^2 = 25$

ANSWERS

1. We'll use the distance formula to find the length of the radius of the circle.

   $d = \sqrt{(x_2 - x_1)^2 + (y_2 - y_1)^2} = \sqrt{(-1 - 2)^2 + (3 - 1)^2} = \sqrt{(-3)^2 + (2)^2} = \sqrt{9 + 4} = \sqrt{13}$

2. The center point is $(-2,-1)$. In order for us to write the equation of the circle, we have to find its radius by using the distance formula.

   $d = \sqrt{(x_2 - x_1)^2 + (y_2 - y_1)^2} = \sqrt{(-2 - -2)^2 + (-1 - -3)^2} = \sqrt{(-2 + 2)^2 + (-1 + 3)^2} = \sqrt{0 + 4} = \sqrt{4} = 2$

   The radius is 2.

   Substitute the values $h = -2$, $k = -1$, and $r = 2$ into the equation of a circle.

   $(x - h)^2 + (y - k)^2 = r^2$

   $(x - -2)^2 + (y - -1)^2 = 2^2$

   $(x + 2)^2 + (y + 1)^2 = 4$

3. The standard form of a circle is $(x - h)^2 + (y - k)^2 = r^2$.

The circle we were given, $x^2 + y^2 = 1$ can be written as $(x - 0)^2 + (y - 0)^2 = 1^2$; therefore, its center point is (0,0) and its radius is 1. When we label the four points on the circumference of this circle, we move 1 unit to the right, left, above, and below the origin, which is the center point. This equation and its graph should look familiar. It's the unit circle.

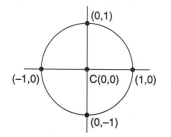

4. The center point is (4,−5), and the radius is 5. To find four points on the circumference, we have to move 5 units to the right, left, above, and below (4,−5).

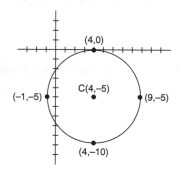

Now that you know how to graph a circle and how to write the equation of a circle given its center point and radius, let's take another look at the standard equation of a circle.

$(x - h)^2 + (y - k)^2 = r^2$	Let's see what happens if we expand this equation.
$(x - h)(x - h) + (y - k)(y - k) = r^2$	Distribute.
$x^2 - hx - hx + h^2 + y^2 - ky - ky + k^2 = r^2$	Combine like terms.
$x^2 - 2hx + h^2 + y^2 - 2ky + k^2 = r^2$	This is another form of the equation of the circle.

When it's written in this form, it's called the general form of the equation of a circle.

General form of a circle:

$$Ax^2 + By^2 + Cx + Dy = E$$

When the equation of a circle is written in general form, we can't know its center point and radius just by looking at the equation. For us to find the center and radius of the circle, we have to convert the equation from general form to standard form. In order for us to do that, we have to use a process called completing the square. This process allows us to factor the equation into two perfect squares. One of the perfect squares will be formed from the terms that contain an x; the other will be formed from the terms that contain a y. As we work our way through the following example, we'll explain the process for completing the square.

Steps to complete the square:

Step 1. Check to see if the coefficients of x^2 and y^2 are 1; if not, divide every term of the equation by the coefficient of x^2 and y^2.

Step 2. Group the terms containing an x together in descending order; do the same for the terms containing a y, and move the constant to the other side of the equal sign. (This will reverse the sign of the constant.)

Step 3. Multiply the coefficient of x by $\dfrac{1}{2}$ or divide it by 2, then square that number and add it to both sides of the equation. Do the same for the coefficient of y.

Step 4. Factor. Note that the factored form turns out to use the coefficients of x and y divided by 2.

Example 31:

Find the center point and radius of the following circle.

$$x^2 + y^2 + 2x - 4y - 11 = 0$$

Solution:

When a circle is written in general form, such as $x^2 + y^2 + 2x - 4y - 11 = 0$, it's not possible to see what the center point and the radius are by looking at the equation. We have to complete the square to write the equation of the circle in standard form.

Step 1. Check to see if the coefficients of x^2 and y^2 are 1; if not, divide every term of the equation by the coefficient of x^2 and y^2.

The coefficient in $x^2 + y^2 + 2x - 4y - 11 = 0$ is 1 already so we don't have to divide.

Step 2.   Group the terms containing an x together in descending order; do the same for the terms containing a y, and move the constant to the other side of the equal sign. (Remember to reverse the sign of the constant.)

$(x^2 + 2x) + (y^2 - 4y) = 11$

Step 3.   Multiply the coefficient of x by $\dfrac{1}{2}$ or divide it by 2, then square that number and add it to both sides of the equation. Do the same for the coefficient of y.

The coefficient of x is 2. $2 \div 2 = 1$, $(1)^2 = 1$. Add 1 to both sides of the equation.

The coefficient of y is -4. $-4 \div 2 = -2$, $(-2)^2 = 4$. Add 4 to both sides of the equation.

$(x^2 + 2x + 1) + (y^2 - 4y + 4) = 11 + 1 + 4$

Step 4.   Factor. Note that the factored form turns out to use the coefficient of x and y divided by 2.

$(x^2 + 2x + 1) + (y^2 - 4y + 4) = 16$

$(x + 1)(x + 1) + (y - 2)(y - 2) = 16$

$(x + 1)^2 + (y - 2)^2 = 16$

We have taken a circle written in standard form, $x^2 + y^2 + 2x - 4y - 11 = 0$, and used the process of completing the square to transform it to general form, $(x + 1)^2 + (y - 2)^2 = 16$. Now we can see that the center point of the circle is $(-1,2)$ and the radius is 4.

Example 32:

Find the center and radius of the following circle.

$2x^2 + 2y^2 + 4x + 3y - 3 = 0$

Solution:

Step 1.   Check to see if the coefficient of x^2 is a 1; if not, divide every term of the equation by the coefficient of x^2.

The coefficient of x^2 is 2, so we have to divide every term by 2.

Step 2.   Group the terms containing an x together in descending order; do the same for the terms containing a y, and move the constant to the other side of the equal sign.

$(x^2 + 2x) + \left(y^2 + \dfrac{3}{2}y\right) = \dfrac{3}{2}$

Step 3.   Multiply the coefficient of x by $\dfrac{1}{2}$ or divide it by 2, then square that number, and add it to both sides of the equation. Do the same for the coefficient of y.

$$\frac{1}{2}\left(\frac{2}{1}\right)=1 \quad 1^2=1 \qquad\qquad \frac{1}{2}\left(\frac{3}{2}\right)=\frac{3}{4} \quad \left(\frac{3}{4}\right)^2=\frac{9}{16}$$

$$(x^2+2x+1)+\left(y^2+\frac{3}{2}y+\frac{9}{16}\right)=\frac{3}{2}+1+\frac{9}{16}, \qquad \frac{3}{2}+1+\frac{9}{16}=\frac{24}{16}+\frac{16}{16}+\frac{9}{16}=\frac{49}{16}$$

Step 4.   Factor. Note that the factored form turns out to use the coefficient of x and y divided by 2.

$$(x^2+2x+1)+\left(y^2+\frac{3}{2}y+\frac{9}{16}\right)=\frac{49}{16}$$

$$(x+1)(x+1)+\left(y+\frac{3}{4}\right)\left(y+\frac{3}{4}\right)=\frac{49}{16}$$

$$(x+1)^2+\left(y+\frac{3}{4}\right)^2=\frac{49}{16}$$

Now that we have used the process of completing the square to write the equation of the circle in general form, we can easily see that the center point of the circle is $\left(-1,-\dfrac{3}{4}\right)$ and the radius is $\sqrt{\dfrac{49}{16}}$, which is $\dfrac{7}{4}$.

Example 33:

Find the center point and radius of the following circle. What's special about this circle?

$x^2-6x+y^2-6y+18=0$

Solution:

Step 1.   Check to see if the coefficient of x^2 is 1; if not, divide every term of the equation by the coefficient of x^2.

The coefficients of x^2 and y^2 are already 1, so we don't have to divide.

Step 2.   Group the terms containing an x together in descending order; do the same for the terms containing a y, and move the constant to the other side of the equal sign.

$(x^2-6x)+(y^2-6y)=-18$

Step 3. Multiply the coefficient of x by $\dfrac{1}{2}$ or divide it by 2, then square that number and add it to both sides of the equation. Do the same for the coefficient of y.

$$\frac{1}{2}\left(\frac{-6}{1}\right) = -3 \quad (-3)^2 = 9$$

$$(x^2 - 6x + 9) + (y^2 - 6y + 9) = -18 + 9 + 9$$

Step 4. Factor. Note that the factored form turns out to use the coefficient of x and y divided by 2.

$$(x - 3)^2 + (y - 3)^2 = 0$$

The center point is (3,3) and the radius is 0. What do you think a radius of 0 means? It means $x^2 - 6x + y^2 - 6y + 18 = 0$ is not a circle; it's the point (3,3).

SELF-TEST 7

1. Write this circle in general form.

    $(x - 1)^2 + (y + 2)^2 = 25$

2. Write the equation in general form of the circle with center point (−1,5) and radius 4.

3. Find the center point and radius of the following circles.

    a. $x^2 + y^2 + 8x + 2y = -13$ b. $2x^2 + 2y^2 - 6x + 10y - 1 = 0$

    c. $9x^2 + 9y^2 - 12y - 5 = 0$ d. $x^2 - 6x + y^2 + 10y + 34 = 0$

ANSWERS

1. $(x - 1)^2 + (y + 2)^2 = 25$

    $(x - 1)(x - 1) + (y + 2)(y + 2) = 25$

    $x^2 - x - x + 1 + y^2 + 2y + 2y + 4 = 25$

    $x^2 - 2x + 1 + y^2 + 4y + 4 = 25$

    $x^2 + y^2 - 2x + 4y = 20$

2. $(x + 1)^2 + (y - 5)^2 = 4^2$

    $(x + 1)(x + 1) + (y - 5)(y - 5) = 16$

    $x^2 + x + x + 1 + y^2 - 5y - 5y + 25 = 16$

    $x^2 + y^2 + 2x - 10y = -20$

3. a. $x^2 + y^2 + 8x + 2y = -13$

$(x^2 + 8x) + (y^2 + 2y) = -13$

$(x^2 + 8x + 16) + (y^2 + 2y + 1) = -13 + 16 + 1$

$(x + 4)^2 + (y + 1)^2 = 4$

$C(-4,-1)$ $r = 2$

b. $2x^2 + 2y^2 - 6x + 10y - 1 = 0$

$x^2 + y^2 - 3x + 5y - \dfrac{1}{2} = 0$

$(x^2 + 3x) + (y^2 + 5y) = \dfrac{1}{2}$

$\left(x^2 - 3x + \dfrac{9}{4}\right) + \left(y^2 + 5y + \dfrac{25}{4}\right) = \dfrac{1}{2} + \dfrac{9}{4} + \dfrac{25}{4}$

$\left(x - \dfrac{3}{2}\right)^2 + \left(y + \dfrac{5}{2}\right)^2 = 9$

$C\left(\dfrac{3}{2}, -\dfrac{5}{2}\right)$ $r = 3$

c. $9x^2 + 9y^2 - 12y - 5 = 0$

$x^2 + y^2 - \dfrac{4}{3}y - \dfrac{5}{9} = 0$

$x^2 + \left(y^2 - \dfrac{4}{3}y\right) = \dfrac{5}{9}$

$x^2 + \left(y^2 - \dfrac{4}{3}y + \dfrac{4}{9}\right) = \dfrac{5}{9} + \dfrac{4}{9}$

$(x - 0)^2 + \left(y - \dfrac{2}{3}\right)^2 = 1$

$C\left(0, \dfrac{2}{3}\right)$ $r = 1$

d. $x^2 - 6x + y^2 + 10y + 34 = 0$

$(x^2 - 6x) + (y^2 + 10y) = -34$

$(x^2 - 6x + 9) + (y^2 + 10y + 25) = -34 + 9 + 25$

$(x^2 - 3)^2 + (y^2 + 5)^2 = 0$

$C(3,-5)$ $r = 0$

The radius is 0; therefore, this is not a circle, it's a point.

Applications

Now comes the fun part—when we get to apply the mathematical techniques we have mastered in this chapter to real–life situations. Again, we apologize if some of these applications seem somewhat contrived. The aim is to reinforce the skills you've acquired. In other words, "Use it or lose it."

Example 34:

What is the circumference of this clock?

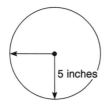

Solution:

$C = 2\pi r = 2(3.14)(5) = 31.4$ inches

Example 35:

Engineers want to build a tunnel through a mountain. The circular distance around the mountain is 2 miles. How long would the tunnel be?

Solution:

The circumference of the circle is 2 miles. The length of the tunnel is the distance through it, which is the diameter of the circle.

$C = \pi d$	Substitute 2 for C and solve for *d*.
$2 = 3.14d$	Divide both sides of the equation by 3.14.
$d = 0.64$ miles	

Example 36:

What is the circumference of a cake that is baked in a pan with a 4-inch radius? What's the area of the top of the cake?

Solution:

$C = 2\pi r = 2(3.14)(4) = 25.12$ inches
$A = \pi r^2 = 3.14(4)^2 = 50.24$ square inches

Example 37:

At football practice, the quarterback threw the ball across the center of a circular field. If the field has a circumference of 167 yards, how far did he throw the ball?

Solution:

$C = \pi d$	Substitute 167 for C and solve for d.
$167 = 3.14d$	Divide both sides of the equation by 3.14.
$d \approx 53.18$ yards	

Example 38:

A circular swimming pool has a diameter of 12 feet. What is its area?

Solution:

The formula for area uses the radius, not the diameter, of a circle. So we'll divide the diameter by 2 to find the radius.

$r = \dfrac{d}{2} = \dfrac{12}{2} = 6$ feet

$A = \pi r^2$ Substitute 6 for the radius.

$A = 3.14(6)^2 = 113.04$ square feet

Example 39:

A circular tarpaulin has an area of 206 square feet. What is its radius?

Solution:

$A = \pi r^2$	Substitute 206 for A and solve for r.
$206 = 3.14r^2$	Divide both sides by 3.14.
$r^2 = \dfrac{206}{3.14}$	Take the square root of both sides of the equation.

$r = \sqrt{\dfrac{206}{3.14}} \approx 8.1$ feet

Example 40:

What is the area of a semicircular piece of land if its diameter is 58 feet?

Solution:

First we have to find the radius of the piece of land. Because a semicircle is half a circle, we'll have to multiply the formula for the area of a circle by 0.5.

$r = \dfrac{d}{2} = \dfrac{58}{2} = 29$ feet

$A = 0.5\pi r^2 = 0.5(3.14)(29)^2 = 1{,}320.37$ square feet

Example 41:

A wedge was cut from the following circular piece of linoleum. What's the area of the wedge?

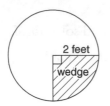

Solution:

The wedge is one-quarter of the circle, so we'll have to adjust the formula for the area of a circle to fit this problem.

$$A = \frac{1}{4}\pi r^2 = \frac{1}{4}(3.14)(2)^2 = 3.14 \text{ square feet}$$

Example 42:

In terms of area, how much larger is a circle with a diameter of 16 inches than a circle with a diameter of 8 inches?

Solution:

We'll find the area of both circles and subtract.

$$A = \pi r^2 = 3.14(8)^2 = 200.96 \text{ square inches}$$
$$A = \pi r^2 = 3.14(4)^2 = 50.24 \text{ square inches}$$

The larger circle is $200.96 - 50.24 = 150.72$ square inches larger than the smaller circle. We can also say the circle with a diameter of 16 inches has an area four times that of the circle with the diameter of 8 inches.

Example 43:

How much would it cost to carpet a circular room with a radius of 30 feet, if the cost of the carpeting is $24.35 per square foot?

Solution:

First we have to find the area, then multiply it by the cost per square foot.

$$\text{Area} = \pi r^2 = 3.14(30)^2 = 3.14(900) = 2{,}826 \text{ square feet}$$

$$\text{Cost} = (\text{number of square feet})(\text{cost per square foot}) = 2{,}826(24.35) = \$68{,}813.10$$

Example 44:

A circular path 4 feet wide is built around a circular garden with a radius of 15 feet. Find the area of the circular path.

Solution:

We'll find the area of the garden and the path. Then we'll find the area of the garden. Last, we'll subtract the area of the garden from the area of the garden and the path. The radius of the garden and the path is 19 (4 + 15).

$A = \pi r^2 = 3.14(19)^2 = 1{,}133.54$ square feet

$A = \pi r^2 = 3.14(15)^2 = 706.5$ square feet

The area of the circular path is 427.04 square feet (1,133.54 − 706.5).

SELF-TEST 7

1. What is the circumference of the following pie?

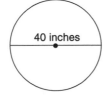

40 inches

2. A circular steel pipe has a circumference of 78.5 inches. What is its diameter?

3. How far would a lumberjack need to chop through the center of a circular tree trunk with a circumference of 109.9 inches?

4. A straight path through a forest is 3 miles long. How long is a semicircular path around the forest?

5. A circular pond has a diameter of 300 feet. Find its area.

6. A merry-go-round has a radius of 18 feet. What is its area?

7. A one-story circular building has an area of 31,400 square feet. What is its diameter?

8. A circular room has a diameter of 15 yards. What would is cost to cover it with wall-to-wall carpeting that costs $12.99 per square yard?

9. A pizza is cut into 8 equal slices. If the radius of the pie is 12 inches, what is the area of 2 of the slices?

10. How much larger is the area of a circular swimming pool with a diameter of 30 feet than of one with a diameter of 10 feet?

11. A circular path 3 feet wide encloses a circular pool. The area covered by the path and the pool is 314 square feet. What's the radius of the pool?

ANSWERS

1. $C = d\pi = 40(3.14) = 125.6$ inches

2. $C = d\pi$

   $78.5 = d(3.14)$

   $d = \dfrac{78.5}{3.14} = 25$ inches

3. $C = d\pi$

   $109.9 = d(3.14)$

   $d = \dfrac{109.9}{3.13} = 35$ inches

4. $\dfrac{1}{2}C = \dfrac{1}{2}d\pi$

   $\dfrac{1}{2}C = \dfrac{1}{2}(3)(3.14)$

   $C = 9.42$ miles

5. $A = \pi r^2 = 3.14(150)^2 = 70{,}650$ square feet

6. $A = \pi r^2 = 3.14(18)^2 = 1{,}017.36$ square feet

7. $A = 3.14r^2$

   $\dfrac{31{,}400}{3.14} = r^2$

   $10{,}000 = r^2$

   $d^2 = \dfrac{125{,}600}{3.14}$

   $100 \text{ feet} = r^2,$

   $d = 200$ feet

8. $A = 3.14\left(\dfrac{15}{2}\right)^2 = 176.625$

   Cost = \$12.99 times area = $12.99(176.625) = \$2{,}294.36$

9. Two slices is one-quarter of the pie.

   $\dfrac{1}{4}A = \dfrac{1}{4}(3.14)(12)^2 = 113.04$ square inches

10. $A = \pi r^2 = 3.14(15)^2 = 706.5$ square feet

$A = \pi r^2 = 3.14(5)^2 = 78.5$ square feet

$706.5 - 78.5 = 628$ square feet larger

The ratio is $\dfrac{706.5}{78.5} = \dfrac{9}{1}$

11. $A = \pi r^2$

$314 = 3.14(r + 3)^2$

$\dfrac{314}{3.14} = (r + 3)^2$

$100 = (r + 3)^2$

$\sqrt{100} = \sqrt{(r + 3)^2}$

$10 = r + 3$

$r = 7$ feet

4 Perimeter and Area of Two-dimensional Polygons

In this chapter we'll work on some problems involving the area and perimeter of complex figures. We'll also work on problems involving everyday applications, such as tiling around a pool or covering a yard with artificial turf. Try the following pretest. If you get all of the problems correct, move on to the next chapter. By the end of this chapter, you should be able to:

- divide a complex figure into smaller sections to find the area and perimeter

- convert from square feet to square yards

- find not just the area or perimeter, but also the cost involved in a given real-world problem

So stop wasting time and get started!

PRETEST

1. What is the perimeter and area of a square with sides 3 inches long?

2. What is the perimeter and area of a rectangle with length 6 and width 2?

3. Find the perimeter and area of a rectangle if the altitude is 5 and the diagonal is 13.

4. Find the perimeter and area of the following isosceles trapezoid.

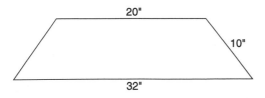

5. Find the perimeter and area of a parallelogram with a base of 15 inches and a height of 6 inches.

6. The following figure contains a square within a rectangle. Find the perimeter and area of the shaded region.

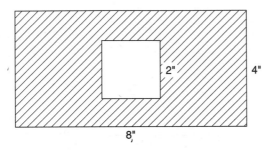

7. The following figure shows a circle in a square.

   a. How much longer is the perimeter of the square than the circumference of the circle?

   b. How much larger is the area of the square than the area of the circle?

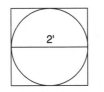

8. The following figure shows a triangle inscribed in a rectangle.

   a. Find the perimeter of the region bounded by ACDE.

   b. Find the area bounded by ACDE.

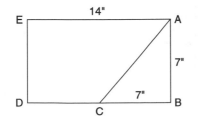

9. Find the perimeter and area of the shape bounded by DEBC in the following figure.

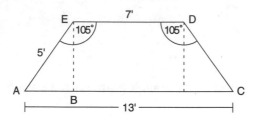

10. Find the perimeter and area of the following geometric shape bounded by BCDE.

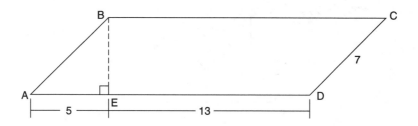

11. Find the perimeter and area of the following complex figure.

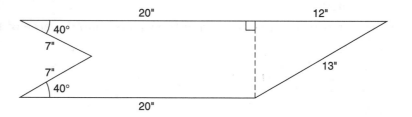

12. You need to place artificial turf in your rectangular yard around your circular pool. If artificial turf costs $12 a square yard, how much would it cost to cover the yard shown in the following figure?

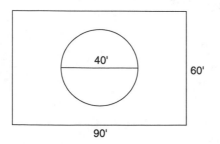

ANSWERS

1. Perimeter = $4s$ = 4(3) = 12 inches

   Area = s^2 = $(3)^2$ = 9 square inches

2. Perimeter = 2(length) + 2(width) = 2(6) + 2(2) = 12 + 4 = 16

   Area = (length)(width) = (6)(2) = 12

3. $a^2 + b^2 = c^2$

   $5^2 + b^2 = 13^2$

   $25 + b^2 = 169$

   $b^2 = 144$

   $b = \sqrt{144} = 12$

   Perimeter = 2(length) + 2(width) = 2(12) + 2(5) = 24 + 10 = 34

   Area = (length)(width) = 12(5) = 60

4. Perimeter = 20 + 10 + 32 + 10 = 72 inches

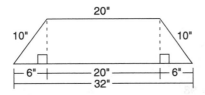

   Area of the rectangle = (length)(width) = (20)(8) = 160 square inches

   $a^2 + b^2 = c^2$

   $6^2 + b^2 = 10^2$

   $36 + b^2 = 100$

   $b^2 = 64$

   $b = \sqrt{64} = 8$ inches

   Area of one triangle is $\frac{1}{2}$(base)(height) = $\frac{1}{2}$(6)(8) = 24 square inches

   Area of both triangles = 48 square inches

   Area of the rectangle = (length)(width) = (20)(8) = 160 square inches

   Area of the trapezoid = 160 + 48 = 208 square inches

5. Perimeter = 2(base) + 2(height) = 2(15) + 2(6) = 30 + 12 = 42 inches

   Area = (base)(height) = (15)(6) = 90 square inches

6. Perimeter = 2(length) + 2(width) = 2(8) + 2(4) = 16 + 8 = 24 inches

   Area of the rectangle = (length)(width) = (8)(4) = 32 square inches

   Area of the square = s^2 = $(2)^2$ = 4 square inches

   Area of the shaded region = 32 square inches − 4 square inches = 28 square inches

7. a. Perimeter of the square = $4s$ = 4(2) = 8 feet

   Circumference of the circle = πd = 3.14(2) = 6.28 feet

   The difference in the perimeter and the circumference is 8 − 6.28 = 1.72 inches.

   b. Area = s^2 = $(2)^2$ = 4 square inches

   Area = πr^2 = $(3.14)(1)^2$ = 3.14 square inches

   The difference in the areas is 4 − 3.14 = 0.86 square inches

8. a. $a^2 + b^2 = c^2$

   $7^2 + 7^2 = (\overline{AC})^2$

   $49 + 49 = (\overline{AC})^2$

   $98 = (\overline{AC})^2$

   $\overline{AC} = \sqrt{98} = 7\sqrt{2} \approx 9.9$

   Perimeter of ACDE = 14 + 9.9 + 7 + 7 = 37.9

   b. Area of the rectangle = 14(7) = 98

   Area of the triangle = $\frac{1}{2}(7)(7)$ = 24.5

   Area of ACDE = 98 − 24.5 = 73.5

9. $a^2 + b^2 = c^2$

   $(\overline{EB})^2 + 3^2 = 5^2$

   $(\overline{EB})^2 + 9 = 25$

   $(\overline{EB})^2 = 16$

   $(\overline{EB}) = \sqrt{16} = 4$ feet

   Perimeter = 7 + 4 + 7 + 3 + 5 = 26 feet

   Area of rectangle = 7(4) = 28 square feet

   Area of triangle = $\frac{1}{2}(3)(4)$ = 6 square feet

   Area of DEBC = 28 + 6 = 34 square feet

10. $a^2 + b^2 = c^2$

$5^2 + b^2 = 7^2$

$25 + b^2 = 49$

$b^2 = 24$

$b = \sqrt{24} \approx 4.9$

Perimeter $= 13 + 7 + 18 + 4.9 = 42.9$

Area of the entire figure $= 18(4.9) = 88.2$

Area of triangle ABE $= \frac{1}{2}(5)(4.9) = 12.25$

Area of figure BCDE $= 88.2 - 12.25 = 75.95$

11. Perimeter $= 20 + 12 + 13 + 20 + 7 + 7 = 79$ inches

To find the area of the figure, we'll begin by finding the area of the triangle on the right side of the figure.

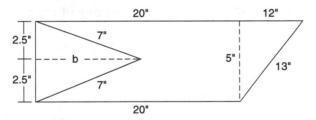

$a^2 + b^2 = c^2$

$12^2 + b^2 = 13^2$

$144 + b^2 = 169$

$b^2 = 25$

$b = \sqrt{25} = 5$ inches

The area of the triangle is $\frac{1}{2}(12)(5) = 30$ square inches.

Next we'll calculate the area of a 20-by-5 rectangle.

Area of the rectangle $= 20(5) = 100$ square inches

The area of the rectangle includes the part to the left that is open, so we'll calculate its area and subtract that from the rectangle.

$a^2 + b^2 = c^2$

$(2.5)^2 + b^2 = 7^2$

$6.25 + b^2 = 49$

$b^2 = 42.75$

$b = \sqrt{42.75} \approx 6.54$

Area $= .5(6.54)(2.5) = 8.175$

Area of the figure $= 100 - 2(8.175) + 30 = 113.65$ square inches

12. Area of the rectangle = (90)(60) = 5,400 square feet

Area of the circle = (3.14)(20)^2 = 1,256 square feet

Area to be covered with artificial turf = 5,400 − 1,256 = 4,144 square feet

Number of square yards = 4,144 ÷ 9 = 460.44 square yards

Cost to cover the area is (460.44)(12) = $5,525.28

So, how did you do? If you got everything right, you may skip the entire chapter. But we recommend that you skim it at least, just to make sure you're able to do everything we're covering. If you got one or two wrong answers, as a minimum you'll need to find the section(s) where this work is covered and go over anything you need to review.

What if you got more than two wrong answers? Well, you know what we're going to tell you. That's right—we want you to work your way through the entire chapter. But if you really know all the material in one of the sections, we'll let you skip it.

Basic Figures

In the last two chapters we talked a lot about the perimeter and area of triangles and circles. Now we'll return to the polygons we reviewed in chapter 1 and find their perimeter and area. We'll begin with the square, followed by the rectangle, the trapezoid, and the parallelogram. But first we'd like to ask you a question: What do these four polygons have in common? Each has four sides. If you knew the answer, maybe you're ready for what comes next.

Squares, rectangles, trapezoids, and parallelograms are all quadrilaterals. And a quadrilateral, as you may recall from chapter 1, is a four-sided polygon.

Squares

As you might well know, *a square contains four right angles (angles of 90°) and has four sides of equal length.*

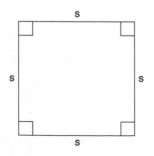

Perimeter of a square formula:

P = 4s, where s stands for the length of a side.

Here's a nice easy one: Find the perimeter of the following square.

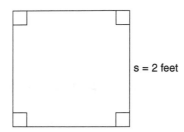

s = 2 feet

P = 4s = 4(2) = 8. The perimeter is 8 feet.

Now *here's* a challenge. Find the area, *A,* of this same square.

Area of a square formula:

A = s²

A = s^2 = 2^2 = 4. The area is 4 square feet.

Rectangles

A rectangle contains four right angles (90°), a pair of equal lengths, and a pair of equal widths.

Perimeter of a rectangle formula:

P = 2l + 2w

Using this formula, find the perimeter of the following rectangle.

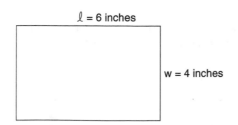

ℓ = 6 inches

w = 4 inches

P = 2l + 2w = 2(4) + 2(6) = 8 + 12 = 20. The perimeter is 20 inches.

Area of a rectangle formula:

A = *lw*

Go ahead and find the area of this rectangle.

A = *lw* = 6(4) = 24. The area is 24 square inches.

Example 1:

Find the area of a rectangle if the length is 15 and the perimeter is 50.

Solution:

To find the area, first we have to find the width. We'll use the formula for perimeter to find the width.

P = 2*l* + 2*w*	Substitute 15 for *l* and 50 for P.
50 = 2(15) + 2*w*	
50 = 30 + 2*w*	Subtract 30 from both sides of the equation.
20 = 2*w*	Divide both sides of the equation by 2.
10 = *w*	The width is 10.

Now that we know the width, we can substitute its value into the formula for area.

A = *lw*

A = 15(10) = 150

Example 2:

Find the perimeter of a rectangle if its width is 9 inches and its area is 117 square inches.

Solution:

To find the perimeter, we have to find the length of the rectangle. We know the area, so we'll start by using the area formula to find the length.

A = *lw*	Substitute 9 for *w* and 117 for A.
117 = 9*l*	Divide both sides of the equation by 9 to find the length.
13 = *l*	The length is 13 inches.

Now that we know the length, it's easy to find the perimeter. Just substitute 13 for *l* and 9 for *w*.

P = 2*l* + 2*w*

P = 2(13) + 2(9) = 26 + 18 = 44 inches

Example 3:
Find the area of a rectangle if its height is 10 and its diagonal is 26.

Solution:
We'll begin by drawing the rectangle.

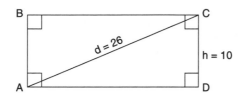

To find the area of this rectangle, we have to find the measure of side *b*. The diagonal divides the rectangle into two right triangles. Because we know the measure of two of the sides of a right triangle, all we have to do is use the Pythagorean theorem to find the measure of the third side of the right triangle. Remember, the Pythagorean theorem states that the sum of the squares of the sides of a right triangle is equal to the square of the hypotenuse. The hypotenuse is the side opposite the right angle.

$b^2 + h^2 = d^2$	Substitute 10 for *h* and 26 for *d*.
$b^2 + (10)^2 = (26)^2$	Square 10 and square 26.
$b^2 + 100 = 676$	Subtract 100 from both sides of the equation.
$b^2 = 576$	Take the square root of both sides of the equation.
$b = 24$	

Now that we know the measure of side *b,* we can find the area of the rectangle.

A = *bh*	*b* is the length of the base and *h* is the height of the triangle.
A = 24(10) = 240	The area is 240.

Trapezoids

A trapezoid is a quadrilateral having two and only two parallel sides. The legs are its non-parallel sides, the bases are its parallel sides.

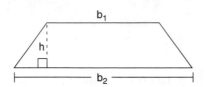

Example 4:

What is the perimeter of the following trapezoid?

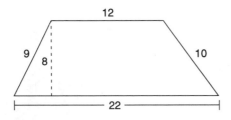

Solution:

P = 9 + 12 + 10 + 22 = 53

The next thing we're going to tackle is finding the area of a trapezoid. *The area of a trapezoid equals one-half the product of its height and the sum of its bases.*

Area of a trapezoid formula:

$$A = \frac{1}{2}h(b_1 + b_2)$$

By convention, b_2 is the longer base and b_1 is the shorter base. As long as we're given or can solve for the height and both bases, we should have no problem finding the area of any trapezoid. Just plug the numbers into the formula and simplify.

Example 5:

Find the area of the trapezoid in example 4.

Solution:

$$A = \frac{1}{2}h(b_1 + b_2)$$ Substitute 8 for h, 22 for b_2 and 12 for b_1.

$$A = \frac{1}{2}(8)(22 + 12) = 4(34) = 136$$

Example 6:

The area of the following trapezoid is 8. Find its perimeter.

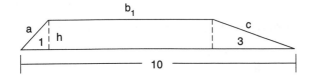

Solution:

To find its perimeter, we have to find the sum of all its sides. First we have to find the length of all the sides that aren't labeled. For reference purposes, let's call the unknown sides a, b_1, and c.

$A = \frac{1}{2}h(b_1 + b_2)$	We already know the area and the measure of b_2.
$8 = \frac{1}{2}h(b_1 + 10)$	This equation has two variables (unknowns) in it. We can't solve for either variable until we find the value for one of them. We don't have enough information to solve for h, but we do have enough information to solve for b_1. $b_1 = 10 - 3 - 1 = 6$.
$8 = \frac{1}{2}h(10 + 6)$	Add the 10 and the 6.
$8 = \frac{1}{2}h(16)$	Multiply $\frac{1}{2}$ times 16.
$8 = 8h$	Divide both sides of the equation by 8.
$h = 1$	

Now that we know the measure of h, we can use the Pythagorean theorem to find the measures of sides a and c.

$a^2 = 1^2 + h^2$	Substitute 1 for h.
$a^2 = 1 + 1$	
$a^2 = 2$	Take the square root of both sides of the equation.
$a = \sqrt{2}$	

Now we'll use the same procedure to find the measure of c.

$c^2 = 1^2 + 3^2$

$c^2 = 1 + 9$

$c^2 = 10$

$c = \sqrt{10}$

The perimeter is $10 + \sqrt{2} + 6 + \sqrt{10} \approx 20.58$.   The $\approx$ means approximately.

Now we're going to find the perimeter and area of an *isosceles trapezoid. An isosceles trapezoid has two legs of equal length.* It also has two pairs of equal angles.

Example 7:

Find the perimeter and area of the following isosceles trapezoid.

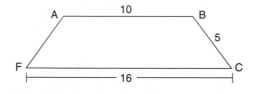

Solution:

Because this is an isosceles trapezoid, we can assume the lengths of sides AF and BC are equal. AF = BC = 5. Now that we know the measure of side AF, all we have to do to find the perimeter of trapezoid ABCF is add the sides.

P = 10 + 5 + 16 + 5 = 36

Now we're going to find the area. But do we have enough information?

No! We don't know the height of this trapezoid. Can you think of how we can find the height? We'd really like you to think about this. If you have any ideas, see if you can come up with an actual number. You'll learn a lot more math by trying to work things out for yourself rather than waiting for us to show you how to do everything.

Did you come up with the height? If you found the height is 4, you're right. Did you use triangles? First we have to drop two vertical dotted lines from the top left corner and the top right corner, forming two right triangles. To find the answer, let's label the point where the vertical line on the left meets the base as E and the same point on the right D.

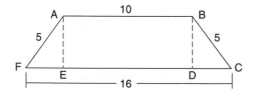

Because these two right triangles have identical angles and the same hypotenuse, △AEF and △BDC are congruent. Therefore, DC ≅ EF. To find the measure of DC and EF, we subtract the measure of AB from FC. 16 − 10 = 6. This leaves us with the measure of EF + DC. Since EF ≅ DC, we equally divide 6 by 2 and get 3, which is the measure of FE and also the measure of DC.

Now we have enough information to find the area of the trapezoid. First we use the Pythagorean theorem to find h, then plug in the values for b_1, b_2, and h into our formula for the area of a trapezoid. Do this, and see if your work matches ours.

$h^2 + 3^2 = 5^2$

$h^2 + 9 = 25$ Subtract 9 from both sides of the equation.

$h^2 = 16$ Take the square root of both sides of the equation.

$h = 4$

$A = \dfrac{1}{2}h(b_1 + b_2)$ Substitute 4 for h, 10 for b_1, and 16 for b_2.

$A = \dfrac{1}{2}(4)(10 + 16)$

$A = 2(26) = 52$

If you'd like to check your work, you can find the area of the rectangle ABDE and add it to the area of $\triangle$BCD and $\triangle$AEF. Try that and see if you still get 52.

Solution:

Area of ABDE = lw   Area of BCD $= \dfrac{1}{2}bh$

$A = 10(4) = 40$ $A = \dfrac{1}{2}(3)(4) = 6$

Because $\triangle$BCD $\cong$ $\triangle$AEF, the area of AEF = 6.

Area of the trapezoid = 40 + 6 + 6 = 52

Example 8:

Find the perimeter and height of the following trapezoid.

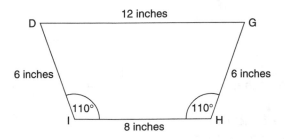

Solution:

Perimeter = 12 + 6 + 8 + 6 = 32 inches.

First we extend a vertical line up from $\angle$I, creating point E, and from $\angle$H, creating point F. This gives us right triangles DEI and GFH. We can assume this is an isosceles trapezoid because its base angles both equal 110°. Since these triangles are congruent, we can assume sides DE and FG are congruent. To find the length of sides DE and FG, all we have to do is subtract the length of side IH (8 inches) from the length of DG (12 inches) and divide the difference by two. DE $\cong$ FG = 2 inches.

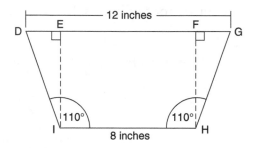

Next we have to find the height of this trapezoid. We'll apply the Pythagorean theorem to one of the right triangles. Let's use right triangle GFH:

$2^2 + h^2 = 6^2$

$4 + h^2 = 36$ Subtract 4 from both sides of the equation.

$h^2 = 32$ Take the square root of both sides of the equation.

$h = \sqrt{32} = \sqrt{16(2)} = 4\sqrt{2}$ inches

Area of DGHI = $\frac{1}{2}h(b_1 + b_2) = \frac{1}{2}(4\sqrt{2})(12 + 8) = (2\sqrt{2})(20) \approx 56.57$ square inches

Let's check our work one last time. (Thereafter, we'll leave it up to you whether to check or not.)

Area of EFHI = lw Area of DEI = $\frac{1}{2}bh$

$A = 8(4\sqrt{2}) \approx 45.25$ square inches $A = \frac{1}{2}(2)(4\sqrt{2}) \approx 5.66$ square inches

Area of DGHI = 45.25 square inches + 5.66 square inches + 5.66 square inches = 56.57 square inches

Parallelograms

A parallelogram is a quadrilateral whose opposite sides are parallel. The symbol we'll use for a parallelogram is □.

Thus, in □ ABCD, AD ∥ BC and AB ∥ DC.
 Now let's find the perimeter of the following parallelogram.

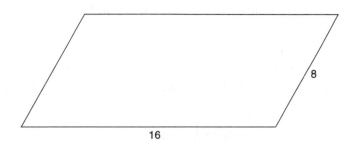

Perimeter = 8 + 16 + 8 + 16 = 48
 The area of a parallelogram equals the product of a side and the height or altitude to that side. If that side happened to be the base, then the formula would be: Area = *bh*.

Example 9:
Find the area of a parallelogram with a base of 10 and a height of 8.

Solution:
A = *bh*
A = 10(8) = 80

Example 10:
Find the area of the following parallelogram.

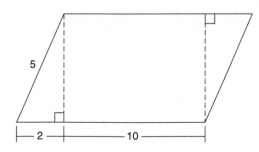

Solution:

In order to find the area of the parallelogram, we have to use the Pythagorean theorem to find its height.

$5^2 = 2^2 + h^2$ — The hypotenuse squared equals the sum of the squares of the sides.

$25 = 4 + h^2$ — Subtract 4 from both sides of the equation.

$21 = h^2$ — Take the square root of both sides of the equation.

$h = \sqrt{21} \approx 4.58$

Now that we know the measure of the height of the parallelogram, all we have to do is substitute the measure of the base and the height into the formula for the area.

$b = 10$ $\qquad\qquad\qquad\qquad h = \sqrt{21} \approx 4.58$

$A = bh = 10(\sqrt{21}) \approx 45.8$

SELF-TEST 1

1. Find the perimeter and the area of a square with a side of 3 inches.

2. If the area of a square is 25, find its perimeter.

3. If the perimeter of a square is 36, find its area.

4. A rectangle has a length of 12 feet and a width of 5 feet. Find its perimeter and area.

5. Find the perimeter and area of a rectangle with a length of 40 yards and a width of 25 yards.

6. Find the area of a rectangle if its length is 20 and its perimeter is 70.

7. If the length of a rectangle is 16 and its area is 64, find the width.

8. Find the perimeter and area of the following trapezoid.

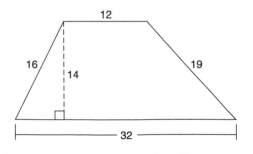

9. Find the perimeter and area of the following trapezoid.

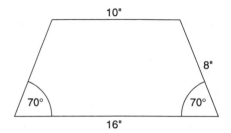

10. Find the perimeter and area of the following trapezoid.

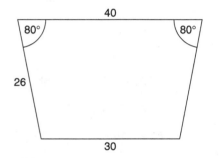

11. One side of a parallelogram is 14 and the base is 20. If the height is 12, find the perimeter and area.

12. Find the area of a parallelogram with a base of 13 inches and a height of 8 inches.

13. Find the area of a rectangle with a diagonal of 13 and a height of 12.

ANSWERS

1. $P = 4s = 4(3 \text{ inches}) = 12$ inches

   $A = s^2 = 3^2 = 9$ square inches

2. $A = s^2$ Substitute 25 for the area.

   $25 = s^2$ Take the square root of both sides of the equation.

   $s = 5$ 5 is the length of each side of the square.

   $P = 4s = 4(5) = 20$ Substitute 5 for *s*. The perimeter of the square is 20.

3. $P = 4s$

   $36 = 4s$

   $9 = s$

   $A = s^2$

   $A = 9^2 = 81$

4. $A = lw$ Substitute 12 for the length and 5 for the width.

   $A = 12(5) = 60$ square feet

   $P = 2l + 2w$

   $P = 2(12) + 2(5) = 24$ feet $+ 10$ feet $= 34$ feet

5. $P = 2l + 2w$ 6. $P = 2 \times$ base $+ 2 \times$ height

   $P = 2(40) + 2(25) = 80 + 50 = 130$ yards $70 = 2(20) + 2 \times$ height

   $A = lw$ $70 = 40 + 2 \times$ height

   $A = 40(25) = 1{,}000$ square yards $30 = 2 \times$ height

   $15 =$ height

   Area $=$ base $\times$ height

   $A = 20 \times 15 = 300$

7. Area $=$ length $\times$ width   8. $P = 16 + 12 + 19 + 32 = 79$

   $64 = 16 \times$ width $A = \frac{1}{2}h(b_1 + b_2) = \frac{1}{2}(14)(32 + 12) = 7(44) = 308$

   $\frac{64}{16} =$ width

   $4 =$ width

9. The base angles are equal. This figure is an isosceles trapezoid.

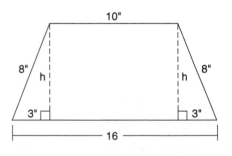

   $h^2 + 3^2 = 8^2$

   $h^2 + 9 = 64$ Subtract 9 from both sides of the equation.

   $h^2 = 55$ Take the square root of both sides of the equation.

   $h = \sqrt{55} \approx 7.42$

   Area $= \frac{1}{2}h(b_1 + b_2) = \frac{1}{2}(\sqrt{55})(16 + 10) \approx 96.41$ square inches

   Perimeter $= 8 + 10 + 8 + 16 = 42$ inches

10.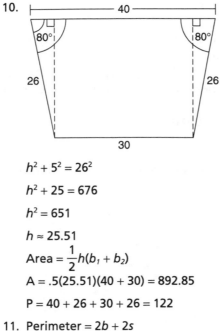

$h^2 + 5^2 = 26^2$

$h^2 + 25 = 676$ Subtract 25 from both sides of the equation.

$h^2 = 651$ Take the square root of both sides of the equation.

$h \approx 25.51$

Area $= \dfrac{1}{2}h(b_1 + b_2)$

$A = .5(25.51)(40 + 30) = 892.85$

$P = 40 + 26 + 30 + 26 = 122$

11. Perimeter $= 2b + 2s$

$P = 2(20) + 2(14) = 40 + 28 = 68$

Area $= bh$

$A = 20(12) = 240$

12. $A = 13(8) = 104$ square inches

13. Area $= bh$

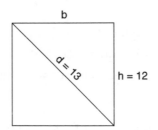

$b^2 + h^2 = d^2$

$b^2 + (12)^2 = (13)^2$

$b^2 + 144 = 169$ Subtract 144 from both sides of the equation.

$b^2 = 25$ Take the square root of both sides of the equation.

$b = 5$

Area $= bh$

$A = 5(12) = 60$

Complex Figures

So far in this book we've discussed circles, triangles, and four types of quadrilaterals: squares, rectangles, trapezoids, and parallelograms. These planar polygons can be combined, as, for example, a square inside a rectangle or a triangle inside a parallelogram. Let's continue to find the perimeters and areas of these complex figures.

Complex Figures Containing Squares

Example 1:

A square with a side of 2 feet is centered inside a square with a side of 4 feet. Find the shaded area in square feet.

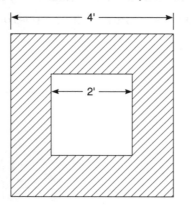

Solution:

All we need to do is find the area of the smaller square and subtract that from the area of the larger square.

Area of smaller square = s^2 = (2 feet)^2 = 4 square feet

Area of the larger square = s^2 = (4 feet)^2 = 16 square feet

Area of the shaded region = 16 square feet − 4 square feet = 12 square feet

Example 2:

In the following figure, we have a square in a corner of a rectangle. Find the perimeter of the shaded area.

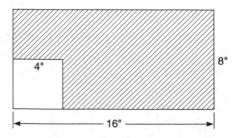

Solution:

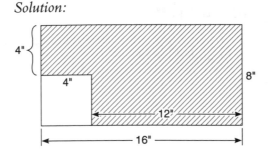

Perimeter = 4 + 16 + 8 + 12 + 4 + 4 = 48

We're on a roll now. See if you can find the shaded area of the region in square inches.

Area of the rectangle = lw = 16 inches × 8 inches = 128 square inches

Area of the square = s^2 = (4 inches)^2 = 16 square inches

Shaded area = 128 square inches − 16 square inches = 112 square inches

Example 3:

Here we have a circle in a square. How much longer is the perimeter of the square than the circumference of the circle? How much larger is the area of the square than the area of the circle?

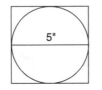

Solution:

Perimeter of the square = $4s$ = 4(5 inches) = 20 inches

Circumference = πd = 3.14(5 inches) = 15.7 inches

P − C = 20 inches − 15.7 inches = 4.3 inches

Now let's see how much larger the area of the square is than the area of the circle. Do that now, and then see if your work matches ours.

Area of the square = s^2 = (5 inches)^2 = 25 square inches

Area of the circle = πr^2 = 3.14(2.5 inches)^2 = 3.14(6.25 sq. in.) = 19.625 square inches

$A_S − A_C$ = 25 square inches − 19.625 square inches = 5.375 square inches

Complex Figures Containing Rectangles

In the following figure we have a triangle inscribed in a rectangle. See if you can find the perimeter of the region bounded by ACDE.

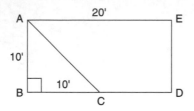

First we'll need to find the length of line segment $\overline{AC}$, which is also the hypotenuse of $\triangle ABC$. Use the Pythagorean theorem:

$$10^2 + 10^2 = c^2$$

$$100 + 100 = c^2$$

$$200 = c^2$$

$$c = \sqrt{200} \approx 14.14 \text{ feet}$$

$\overline{AC} = 14.14$ feet, $\overline{CD} = 10$ feet, $\overline{DE} = 10$ feet, and $\overline{AE} = 20$ feet. The perimeter of ACDE is 54.14 feet.

Now see if you can find the area of the region bounded by ACDE.
We need to subtract the area of $\triangle ABC$ from the area of rectangle ABDE.

Area of ABDE = lw Area of $\triangle ABC = \dfrac{1}{2}bh$

Area = 10 feet (20 feet) = 200 square feet   Area $= \dfrac{1}{2}(10)(10) = 50$ square feet

Area of the rectangle ABDE = Area of triangle ABC = 200 square feet − 50 square feet = 150 square feet

Example 4:
Find the area of the rectangle not included in the area of the circle.

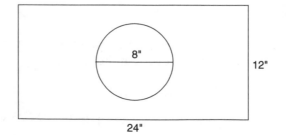

Solution:

Area of the rectangle $= lw = 24$ inches $(12$ inches$) = 288$ square inches

Area of the circle $= \pi r^2 = 3.14$ $(4$ inches$)^2 = 50.14$ square inches

$A_R - A_C = 288$ square inches $- 50.24$ square inches $= 237.76$ square inches

Complex Trapezoids

A trapezoid is formed by a rectangle and a pair of right triangles. We've already worked with complex trapezoids earlier in this chapter.

Example 5:

Find the area of the region bounded by ACDE.

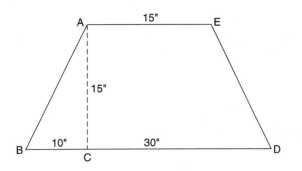

Solution:

Subtract the area of $\triangle$ABC from the area of trapezoid ABDE.

Area of a trapezoid = one-half the height times the sum of the bases

$$A = \frac{1}{2}(b_1 + b_2)$$

Area of trapezoid ABDE $= \dfrac{1}{2}(15$ inches$)(40$ inches $+ 15$ inches$)$

$$= 0.5(15 \text{ inches})(55 \text{ inches})$$

$$= 412.5 \text{ square inches}$$

Area of a triangle = one-half the base times the height

$$A = \frac{1}{2}bh$$

Area of $\triangle$ABC $= \dfrac{1}{2}(10$ inches$)(15$ inches$)$

$$= 0.5(150 \text{ square inches})$$

$$= 75 \text{ square inches}$$

$A_{TRAP} - A_{TRI} = 412.5$ square inches $- 75$ square inches $= 337.5$ square inches

Example 6:

Find the perimeter and area of the shape bounded by DEBC.

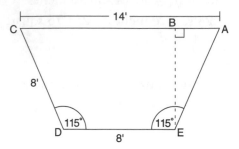

Solution:

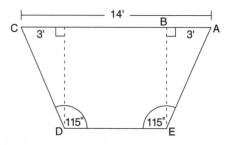

We can easily fill in the value for the length of $\overline{AB}$. Subtract the length of $\overline{DE}$ (8 feet) from the length of $\overline{AC}$ (14 feet) and divide by 2. $(14 - 8) \div 2 = 3$. The length of $\overline{AB}$ is 3 feet. The base angles of this trapezoid are both 115°, so we can assume sides CD and AE are both 8 feet in length. Now all we need to find the perimeter of DEBC is the length of $\overline{BE}$. To find the length of $\overline{BE}$ we'll use the Pythagorean theorem.

$a^2 + b^2 = c^2$	Remember, c is the hypotenuse.
$3^2 + h^2 = 8^2$	Square the 3 and the 8.
$9 + h^2 = 64$	Subtract 9 from both sides of the equation.
$h^2 = 55$	Take the square root of both sides of the equation.
$h = \sqrt{55} \approx 7.42$ feet.	

We can add the sides to find the perimeter of DEBC.

$$P = 8 + 7.42 + 11 + 8 = 34.42 \text{ feet}$$

Now we have to subtract the area of $\triangle ABC$ from the area of the trapezoid AEDC to find the area of the shape bounded by DEBC.

Area of $\triangle ABE = \dfrac{1}{2}bh = 0.5(3)(7.42) = 11.13$ square feet

Area of trapezoid $AEDC = \dfrac{1}{2}(7.42)(8 + 14) = 81.62$ feet

Area of $DEBC$ = Area of $AEDC$ − Area of $\triangle ABE$

Area of $DEBC = 81.62 − 11.13 = 70.49$ square feet

Complex Parallelograms

Example 7:

Find the perimeter and area of the geometric shape bounded by BCDE.

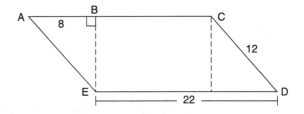

Solution:

First we'll have to find the length of line segment $\overline{BE}$, then the area of ▱ ACDE and the area of $\triangle ABC$.

    The length of $\overline{AE}$ is 12 because the length of $\overline{CD}$ is 12, and we know opposite sides of a parallelogram are equal. To find the length of $\overline{BE}$, we'll use the Pythagorean theorem.

$a^2 + b^2 = c^2$	Substitute 8 for *a* and 12 for *c*.
$8^2 + b^2 = 12^2$	Square 8 and 12.
$64 + b^2 = 144$	Subtract 64 from both sides of the equation.
$b^2 = 80$	Take the square root of both sides of the equation.
$b = \sqrt{80} \approx 8.94$	

Area of $ACDE = bh = 22(8.94) = 196.68$

Area of $\triangle ABE = \dfrac{1}{2}bh = \dfrac{1}{2}(8)(8.94) = 35.76$

Area of $BCDE = 196.68 − 35.76 = 160.92$

Perimeter of $BCDE = 14 + 12 + 22 + 8.94 = 56.94$

Miscellaneous Complex Figures

Now we'll find the perimeter and area of geometric shapes that defy description. It will be relatively easy to find their perimeters, but to find their areas, we'll have to look within the figures for squares, rectangles, triangles, and other familiar shapes.

Example 9:

Find the perimeter and area of the following geometric shape.

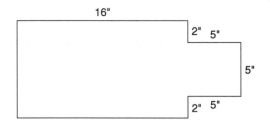

Solution:

You'll notice that this figure is simply a rectangle and a square combined. To find the perimeter, all we have to do is find the left side (or width) of the rectangle, which is 9 inches (2 inches + 5 inches + 2 inches).

Perimeter = 9 + 16 + 2 + 5 + 5 + 5 + 2 + 16 = 60 inches

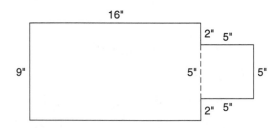

To find the total area, we have to add the area of the square to the area of the rectangle.

The area of the square is $A = s^2 = 5^2 = 25$ square inches.

The area of the rectangle is $A = lw = (16)(9) = 144$ square inches.

Total area = 25 square inches + 144 square inches = 169 square inches

Example 10:

Find the perimeter and area of this complex figure.

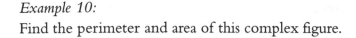

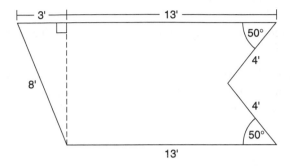

Solution:

Can you recognize three familiar geometric figures here? There is a pair of triangles, and, if you look more closely, there's a rectangle. Adding a vertical dotted line to the right side of the figure makes the rectangle and one of the triangles more apparent.

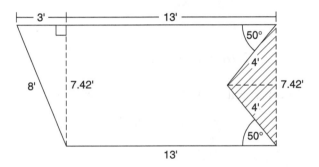

Now let's find the missing leg of the right triangle to the left of the figure.

$a^2 + b^2 = c^2$	Substitute 3 for a and 8 for c.
$3^2 + b^2 = 8^2$	Square the 3 and the 8.
$9 + b^2 = 64$	Subtract 9 from each side of the equation.
$b^2 = 55$	Take the square root of both sides of the equation.
$b = \sqrt{55} \approx 7.42$ feet	

Perimeter = 3 + 13 + 4 + 4 + 13 + 8 = 45 feet

Finding the area is a bit more difficult. We'll add the area of the rectangle to the area of the triangle to the left, then subtract the areas of the shaded pair of triangles to the right.

Area of the rectangle = lw = (13)(7.42) = 96.46 square feet

Area of the left triangle = $\frac{1}{2}bh$ = $\frac{1}{2}$(3)(7.42) = 11.13 square feet

$A_R + A_T$ = 96.46 square feet + 11.13 square feet = 107.59 square feet

Finally, we have to find the areas of the congruent triangles to the right of the complex figure and subtract them from the 107.59-square-foot area we just found. To do that, we need to find the length of the common leg of these congruent triangles.

$a^2 + b^2 = c^2$

$(3.71)^2 + b^2 = 4^2$

$13.76 + b^2 = 16$ square feet

$b^2 = 2.24$ square feet

$b = \sqrt{2.24} \approx 1.5$ feet

Area = $\frac{1}{2}bh$ = $\frac{1}{2}$(1.5)(3.71) = 2.78 square feet

The area of the pair of congruent triangles is 2 × 2.78 = 5.56 square feet.

A = 107.59 square feet − 5.56 square feet = 102.03 square feet

SELF-TEST 2

1. Find the perimeter and the area of the geometric shape bounded by ABCDEF.

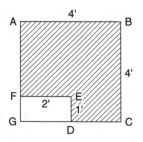

2. Find the perimeter and the area of the geometric figure bounded by ABCEF.

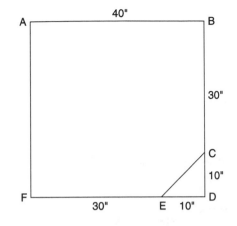

3. Find the perimeter and the area bounded by the geometric shape BCDE.

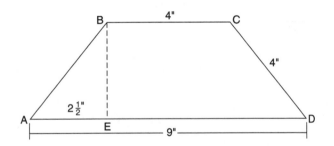

4. Find the perimeter and area bounded by the geometric shape ABCEF.

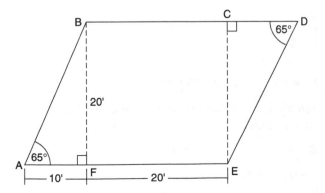

5. Find the perimeter and area of this complex figure.

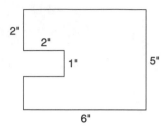

6. Find the perimeter and area of this complex figure.

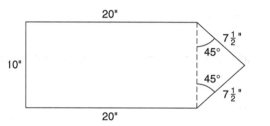

1.

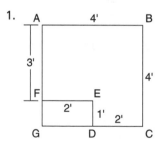

Perimeter = 4 + 4 + 2 + 1 + 2 + 3 = 16 feet

To find the area of figure ABCDEF, we have to subtract the area of rectangle DEFG from the area of square ABCG.

Area of ABCG = s^2 = 4^2 = 16 square feet

Area of DEFG = lw = (2)(1) = 2 square feet

Area of ABCDEF = 16 square feet – 2 square feet = 14 square feet

2.

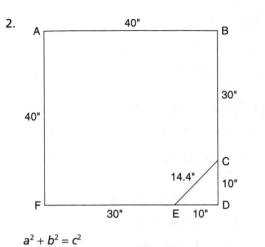

$a^2 + b^2 = c^2$

$10^2 + 10^2 = c^2$ Substitute 10 for *a* and for *b*.

$100 + 100 = c^2$ Square the 10s.

$200 = c^2$ Take the square root of both sides of the equation.

$c = \sqrt{200} \approx 14.14$ inches

Perimeter = 40 + 30 + 14.14 + 30 + 40 = 154.14 inches

To find the area of ABCEF, we have to subtract the area of △CDE from the area of rectangle ABDF.

Area of $\triangle CDE = \frac{1}{2}bh = \frac{1}{2}(10)(10) = \frac{1}{2}(100) = 50$ square inches

Area of rectangle ABDF = lw = (40)(30) = 1,200 square inches

Area of ABCEF = 1,200 square inches − 50 square inches = 1,150 square inches

3.

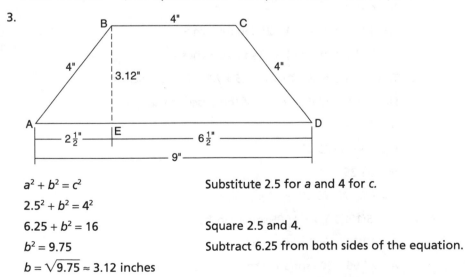

$a^2 + b^2 = c^2$ Substitute 2.5 for *a* and 4 for *c*.

$2.5^2 + b^2 = 4^2$

$6.25 + b^2 = 16$ Square 2.5 and 4.

$b^2 = 9.75$ Subtract 6.25 from both sides of the equation.

$b = \sqrt{9.75} \approx 3.12$ inches

Perimeter = 4 + 4 + 6.5 + 3.12 = 17.62 inches

Area of BCDE = the area of BCDA − the area of △BEA

Area of BCDA = $\frac{1}{2}h(b_1 + b_2) = \frac{1}{2}(3.12)(4 + 9) \approx 20.28$ square inches

Area of △BEA = $\frac{1}{2}bh = \frac{1}{2}(2.5)(3.12) = 3.9$ square inches

Area of BCDE = 20.28 square inches − 3.9 square inches = 16.38 square inches

4. To find the perimeter of this figure, we have to have to find the measure of side AB and side CE.

$a^2 + b^2 = (\overline{AB})^2$ Substitute 20 for a and 10 for b.

$20^2 + 10^2 = (\overline{AB})^2$ Square 20 and 10.

$400 + 100 = (\overline{AB})^2$ Add 400 and 100.

$500 = (\overline{AB})^2$ Take the square root of both sides of the equation.

$\overline{AB} = \sqrt{500} = 10\sqrt{5} \approx 22.36$

$P = 22.36 + 20 + 20 + 20 + 10 = 92.36$ feet

$A = 20^2 + \frac{1}{2}(10)(20) = 400 + 100 = 500$ square feet

5. The perimeter is 6 + 5 + 6 + 2 + 2 + 1 + 2 + 2 = 26 inches.

We're going to list three ways to find the area for this figure. The first calculates the area of a 6 by 5 rectangle and subtracts out the 2 by 1 section.

$A = 6(5) - 1(2) = 28$ square inches

The second and third approaches divide the figure into three smaller rectangles and add their areas.

$A = 4(5) + 2(2) + 2(2) = 28$ square inches

$A = 2(6) + 1(4) + 2(6) = 28$ square inches

6. The perimeter is 10 + 20 + 7.5 + 7.5 + 20 = 65 inches.

First we'll find the height of the isosceles triangle.

$h^2 + 5^2 = (7.5)^2$

$h^2 + 25 = 56.25$

$h^2 = 31.25$

$h = \sqrt{31.25} \approx 5.59$

$A = .5(10)(5.59) = 27.95$ square inches

$A = 10(20) = 200$ square inches

$A = 27.95 + 200$ square inches = 227.95 square inches

Applications

Now we come to the fun part of the chapter, when we get to use all of the great skills we've developed in the first two sections.

Example 11:

You'd like to pave the area around your pool. Your pool is square with a side of 30 feet. The pool is centered in a square area with a side of 40 feet. How many square feet will you have to pave?

Solution:

It helps to draw a picture.

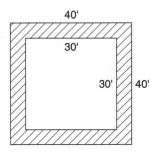

Area paved = total area − area of the pool

Total area = s^2 = $(40)^2$ = 1,600 square feet

Area of the pool = s^2 = $(30)^2$ = 900 square feet

Area paved = 1,600 square feet − 900 square feet = 700 square feet

Example 12:

A circular marble floor with a diameter of 30 feet sits in the center of a rectangular room with a wood floor that is 96 feet by 81 feet. If carpeting cost $14 a square yard, how much would it cost to cover the wooden floor? (Hint: Be sure to convert feet into yards.)

Solution:

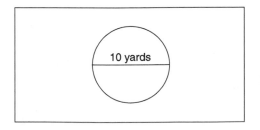

Before we begin to find the area, we have to convert the 96 feet and the 81 feet to yards. We know 3 feet equal 1 yard, so we'll divide 96 by 3 and 81 by 3.

96 feet = 32 yards

81 feet = 27 yards

To find the number of square yards to be carpeted, we have to subtract the area of the marble part of the floor from the total area of the floor.

Area to be carpeted = the area of the floor − area of the marble part of the floor

Area of the floor = (32)(27) = 864 square yards

Area of the marble part of the floor = πr^2 = (3.14)(5)^2 = (3.14)(25) = 78.5 square yards

Area to be carpeted = 864 − 78.5 = 785.5 square yards

Cost of the area to be carpeted = ($14)(785.5) = $10,997.00

Example 13:

A rectangular field is 400 yards wide and 800 yards long. A path goes around the field at a distance of 10 yards from the field. What distance would you walk on the path if you went all the way around the field? (Hint: Draw a picture of the path and the field.)

Solution:

To find the distance we would walk around the field on the path, we have to find the perimeter of the field with the path. Since the path is 10 yards away from every end of the field, the width of the field with the path is 400 yards + 10 yards + 10 yards, which is 420 yards. The length of the field with the path is 800 yards + 10 yards + 10 yards, which is 820 yards. The perimeter of the field with the path is 820 yards + 820 yards + 420 yards + 420 yards, which is 2,480 yards.

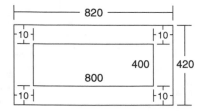

Example 14:

A farmer owns a field that is shaped like a trapezoid, as shown in the following illustration. If he sells off the part of his field represented by △AEF, how many square feet does he have left?

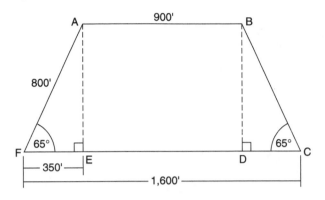

Solution:

To find the area of the remaining figure ABCE, we have to subtract the area of triangle AEF from the area of trapezoid ABCF. The formula for the area of a trapezoid is Area $= \frac{1}{2}$(height)(sum of the bases) $= \frac{1}{2}h(b_1 + b_2)$.

We don't know the height of the trapezoid, so we'll use the Pythagorean theorem to find the height of triangle AEF. The height is side *AE*. We'll call the length of AE side *b*.

$a^2 + b^2 = c^2$	Substitute 350 for *a* and 800 for *c*.
$(350)^2 + b^2 = (800)^2$	Square 350 and 800.
$122{,}500 + b^2 = 640{,}000$	Subtract 122,500 from both sides of the equation.
$b^2 = 517{,}500$	Take the square root of both sides of the equation.
$b = \sqrt{517{,}500} \approx 719.37$	

Now that we know the length of the height AE, we can use the formula $A = \frac{1}{2}h(b_1 + b_2)$.

Area of the trapezoid $= (.5)(719.37)(1600 + 900) = 899{,}212.5$

Area of the triangle $= \frac{1}{2}(base)(height) = \frac{1}{2}(350)(719.37) = 125{,}889.75$

Area of figure ABCE $= 899{,}212.5 - 125{,}889.75 = 773{,}322.75$ square feet

The farmer has 773,322.75 square feet left.

Example 15:

We have drawn the top view (like what you would see from an airplane) of a mansion. The building is basically rectangular and has two square wings. Find the perimeter and the square footage of this mansion.

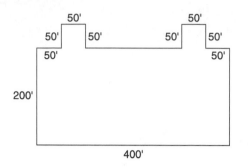

Solution:

To find the perimeter, we begin by finding the measure of the top middle section of the figure. To do this, all we have to do is see that the length of the lower section of the figure between the squares is 400 feet and subtract four sets of 50-foot lengths that are on each side of the length along the upper section and along the wings of the mansion. The length of the upper middle section of the figure is 200 feet.

Perimeter = 200 + 50 + 50 + 50 + 50 + 200 + 50 + 50 + 50 + 50 + 200 + 400 = 1,400 feet

We'll find the area of the mansion by adding the area of the rectangle to the area of the two squares.

Area of the rectangle = length × width = (400)(200) = 80,000 square feet

Area of one wing (square) = s^2 = $(50)^2$ = 2,500 square feet

Area of the mansion = 80,000 + 2,500 + 2,500 = 85,000 square feet

SELF-TEST 3

1. In Pleasantville (the town, not the movie), all the backyards are square. How large a backyard would you need (in square feet) to accommodate a circular pool with a radius of 10 feet?

2. How much is the (a) perimeter and (b) area of the shaded triangular region shown in the following figure? (c) How much would it cost to put a fence around it if fencing material cost $22 per foot?

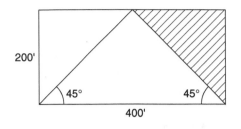

3. The only way to drive from Amityville to Cowtown was through Blahsburgh (see map). But in 2002 a new highway was opened connecting Amityville and Cowtown directly. How many miles would you save on a one-way trip between Amityville and Cowtown? (Hint: First find the distance between Dullsville and Epiphany Springs.)

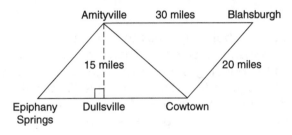

4. Imagine the following map is a field planted with corn.

   a. Find the perimeter of the field.

   b. Find the area of the field.

   c. If you could grow a bushel of corn on every 40 square feet of land, how many bushels could you grow on this field?

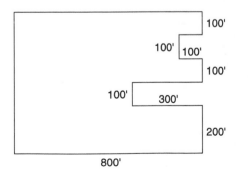

5. Bill Gonzalez needs to buy new carpeting for his living room, which is shown in the following diagram. He won't need to lay carpet in the two circular areas. If carpeting costs $14.99 a square yard, how much will the carpeting cost him?

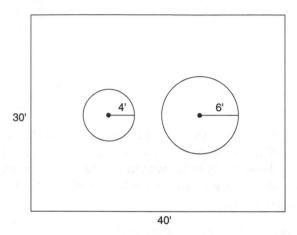

6. The wall of an old office building needs to be whitewashed. Of course, we don't want to whitewash the windows. There are five windows of equal size on each floor, and the building is six stories high. One pail of whitewash costs $5 and can cover 500 square feet. The building is 80 feet high and 50 feet in width. Each window is 6 feet high and 3 feet wide. How much will it cost to whitewash the wall in the following illustration?

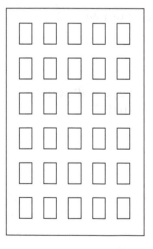

ANSWERS

1.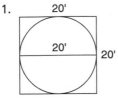

Area of the backyard = s^2 = $(20)^2$ = 400 square feet

2. $a^2 + b^2 = c^2$ Substitute 200 for a and 200 for b.

$200^2 + 200^2 = c^2$ Square the 200s.

$40,000 + 40,000 = c^2$ Add 40,000 and 40,000.

$80,000 = c^2$ Take the square root of both sides of the equation.

$c = \sqrt{80,000} \approx 282.84$ feet

  a. Perimeter = 200 feet + 200 feet + 282.84 feet = 682.84 feet

  b. Area = $\frac{1}{2}bh = \frac{1}{2}(200)(200)$ = 20,000 square feet

  c. Cost of the fence = $22(682.84) = \$15,022.48$

3. We'll begin by finding the distance from Dullsville to Epiphany Springs, then subtract that distance from 30 miles to find the distance between Dullsville and Cowtown. Using that information, we'll be able to find the distance between Amityville and Cowtown.

$a^2 + b^2 = c^2$ Substitute 15 for b and 20 for c.

$a^2 + (15)^2 = (20)^2$ Square 15 and 20.

$a^2 + 225 = 400$ Subtract 225 from each side of the equation.

$a^2 = 175$ Take the square root of both sides of the equation.

$a = \sqrt{175} \approx 13.23$ miles

Distance between Dullsville and Cowtown = 30 miles – 13.23 miles = 16.77 miles

Now we'll find the distance directly from Amityville to Cowtown.

$a^2 + b^2 = c^2$ Substitute 15 for a and 16.77 for b.

$(15)^2 + (16.77)^2 = c^2$ Square 15 and 16.77.

$225 + 281.23 = c^2$ Add 225 and 281.23.

$506.23 = c^2$ Take the square root of both sides of the equation.

$c = \sqrt{506.23} \approx 22.5$ miles

Before the direct route from Amityville to Cowtown, the old route through Blahsburgh had a distance of 50 miles. Using the new route you save 50 – 22.5 = 27.5 miles.

4. a. Perimeter = 800 + 200 + 300 + 100 + 300 + 100 + 100 + 100 + 100 + 100 + 800 + 600 = 3,600 feet

b. We'll solve this problem in two ways. The first involves subtraction, the other addition. To find the area of the field, we'll find the area of a 600 foot by 800 foot lot. Then we'll subtract the areas of the 100 foot by 100 foot square and the area of the 100 by 300 foot rectangle.

Area of the 600 foot by 800 foot lot = (600)(800) = 480,000 square feet

Area of the 100 foot by 100 foot square = $(100)^2$ = 10,000 square feet

Area of the 100 foot by 300 foot rectangle = (100)(300) = 30,000 square feet

Area of the figure = 480,000 − 10,000 − 30,000 = 440,000 square feet

The addition method requires us to subdivide the figure into smaller sections and add their areas.

We're going to divide the figure into a rectangle of dimensions 600 feet by 500 feet (800 − 300). We have two additional rectangles of 100 feet by 300 feet; another rectangle of 200 feet by 300 feet; and, last, a rectangle of 100 feet by 200 feet.

Area of the 600 foot by 500 foot rectangle = (600)(500) = 300,000 square feet

Area of the 200 foot by 300 foot rectangle = (200)(300) = 60,000 square feet

Area of the 100 foot by 100 foot square = (100)(100) = 10,000 square feet

Area of the figure = 300,000 + 60,000 + 2(30,000) + 20,000 = 440,000 square feet

c. $\dfrac{440,000}{40}$ = 11,000 bushels

5. First we'll find the area of the floor to be carpeted by subtracting the areas of the circular regions from the rectangular floor. Then we'll multiply the area by the cost per square yard. Notice the cost is per square yard and the measurements are in terms of feet, not yards. We know 3 feet equal 1 yard, so we'll divide the feet by 3 to convert them to yards.

30 ÷ 3 = 10 yards   40 ÷ 3 = 13.33 yards   4 ÷ 3 = 1.33 yards   6 ÷ 3 = 2 yards

Area of the rectangle = 10(13.33) = 133.3 square yards

Area of the smaller circle = $(3.14)(1.33)^2$ = 5.55 square yards

Area of the larger circle = $(3.14)(2)^2$ = 12.56 square yards

Area to be carpeted = 133.3 − 5.55 − 12.56 = 115.19 square yards

Cost = 115.19($14.99) = $1,726.70

6. First we have to find the area of the wall. Then we'll find the area of 1 window and multiply by 30. Next we'll subtract the area of the 30 windows from the area of the wall to find the area to be whitewashed. We'll divide that area by 500 square feet to find the number of pails of whitewash needed. Finally we'll multiply the number of pails needed by $5 to find out how much it will cost to whitewash the wall.

Area of the wall = (80)(50) = 4,000 square feet

Area of one window = (6)(3) = 18 square feet

Area of 30 windows = (18)(30) = 540 square feet

Area to be whitewashed = 4,000 square feet − 540 square feet = 3,460 square feet

Number of pails of whitewash needed = 3,460 ÷ 500 = 6.92 pails

Cost of the whitewashing the wall is (6.92)($5) = $34.60

5 Volume and Surface Area of Three-dimensional Polygons

Welcome to the third dimension. Until now we've been doing plane geometry, which involves only two dimensions: generally width and length, or base and height. To these we now add depth, as we move from plane geometry to solid geometry.

You're probably much too young to remember three-dimensional (3D) movies, which introduced the third dimension to cinematography. Back in the mid-1950s, moviegoers would be given special disposable dark glasses to wear. The effect was a lot like watching an IMAX movie today (only without the glasses). The most famous 3D movie of the time was *Cinerama,* which provided viewers with the thrilling experience of riding at the front of a roller coaster. Definitely not a movie for the squeamish.

Now back to the less thrilling world of solid geometry. When you have completed this chapter, you will know how to compute the volume and surface area of

- cubes
- rectangular solids
- pyramids
- spheres
- cylinders
- cones
- complex figures

Before we get started, we'd like you to try the following pretest. If you answer all the problems correctly, you can skip to the next chapter.

1. How much larger is the surface area of a cube with an edge of 6 meters than a cube with an edge of 4 meters?

2. Find the volume of a rectangular solid with a length of 8 inches, a width of 6 inches, and a height of 4 inches.

3. What is the surface area of a three-sided pyramid that has a height of 7 feet and a base with an area of 12 feet?

4. If natural gas costs $2 per cubic foot, how much would it cost to fill a spherical tank with a radius of 40 feet?

5. Ike's Ice Cream Parlor serves one-scoop ice cream cones with a perfect hemispherical scoop of ice cream on the cone. If the radius of the cone and hemisphere is 2 inches and the cone has a height of 6 inches, how many cubic inches of ice cream are in Ike's ice cream cones?

6. How much concrete is needed to build a path $\frac{1}{2}$ inch deep, 40 feet long, and 8 feet wide? If the concrete costs $4 per cubic foot, how much will be spent on concrete?

7. How much larger is the volume of a sphere with a radius of 3 feet than a cube with an edge of 2 feet?

8. A sphere with a radius of 10 feet is placed inside a hollow cube with an edge of 20 feet. Water is then pumped into the cube until full. How many cubic feet are pumped into the cube?

9. Rectangular solid A is 16 feet long, 12 feet wide, and 4 feet high. Rectangular solid B is 14 feet long, 14 feet wide, and 3 feet high. We want to carpet the surface areas of A and B, except for their bottoms. Carpeting costs $8.50 per square foot. Does it cost more to carpet A or B? How much more does it cost?

1. The surface area of the first cube is $SA = 6e^2 = 6(6)^2 = 216$ square meters.

   The surface area of the second cube is $SA = 6e^2 = 6(4)^2 = 96$ square meters.

   The difference in the surface areas is 216 square meters − 96 square meters = 120 square meters.

2. $V = lwh = (8)(6)(4) = 192$ cubic inches.

3. $SA = 3\frac{1}{2}Bh = 3\frac{1}{2}(12)(7) = 126$ square feet.

4. $V = \frac{4}{3}\pi r^3 = (1.33)(3.14)(40)^3 = 267{,}277$ cubic inches

   Cost = $2(267,277) = \$534{,}554$

5. $V = \frac{2}{3}\pi r^3 = (.67)(3.14)(2)^3 = 16.8304$ cubic inches

   $V = \frac{1}{3}\pi r^2 h = (.33)(3.14)(2)^2(6) = 24.8688$ cubic inches

   Total volume = 16.8304 cubic inches + 24.8688 cubic inches = 41.6992 cubic inches

6. $V = lwh = \frac{1}{2}(40)(8) = 160$ cubic feet

   Cost = $4(160) = \$640$

7. $V = \frac{4}{3}\pi r^3 = (1.33)(3.14)(3)^3 = 112.757$ cubic feet

   $V = e^3 = (2)^3 = 8$ cubic feet

   The difference in the volumes is 112.757 cubic feet − 8 cubic feet = 104.757 cubic feet.

8. $V = e^3 = (20)^3 = 8{,}000$ cubic feet

   $V = \frac{4}{3}\pi r^3 = (1.33)(3.14)(10)^3 = 4{,}176.2$ cubic feet

   The amount of cubic feet of water pumped into the cube is 8,000 cubic feet − 4,176.2 cubic feet = 3,823.8 cubic feet.

9. Rectangular solid A:

   Top: 16(12) = 192

   Sides: 4(12)(2) = 96

   Front/back: 16(4)(2) = 128

   Area of A = 416 square feet

   Cost to carpet A: $8.50(416) = \$3,536

   Rectangular solid B:

   Top: 14(14) = 196

   Sides: 14(3)(2) = 84

   Front/back: 14(3)(2) = 84

   Area of B = 364 square feet

   Cost to carpet B: $8.50(364) = \$3,094

   $3,536 − $3,094 = $442. It costs $442 more to carpet A than B.

Cubes

A cube is a rectangular solid with equal length, width, and height (see the following figure). Can you think of examples of a cube? There are sugar cubes, dice, office cubes, and even ice cubes (although most ice cubes are not really of equal length, width, and height). Back in the 1930s, cubism was a popular form of abstract art, but today, of course (if you'll pardon the pun), the art world considers cubism square.

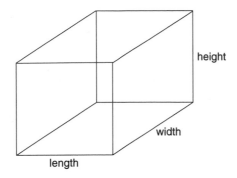

You'll remember from the last chapter that the area of a rectangle is its length times its width—A = *lw*—and that the area of a square is the square of one of its sides, because a square's length and width are always equal: A = s^2. Can you guess at the formula for the volume of a cube?

All the sides of a cube are equal. You could say that the volume of a cube is equal to the cube of any of its sides, or you could say V = *lwh* (that is, length times width times height), which is the same as V = s^3. Both of these formulas are right. But in mathematics, we like to keep introducing new terms to keep you on your toes. One new term we're going to introduce is edge. *An edge of a shape is the line, or one of the lines, defining the outline of that shape.* In a cube, the edges are formed where the defining surfaces meet.

Can you name the surfaces of a cube? (Hint: There are six, four of which have the same name.) The six surfaces of a cube are the four sides, the top, and the bottom. If you turn a cube on its side, then that side becomes the bottom of the cube. It doesn't really matter, because the sides, top, and bottom of cubes have identical dimensions. For instance, if a cube has an edge of 9 inches, the length and width of the top, bottom, and sides are each 9 inches by 9 inches.

Volume of a cube formula is

Volume = e³

where *e* stands for edge

So what's the volume of a cube with an edge of 9 inches?

Volume = e^3 = (9 inches)^3 = 9 inches × 9 inches × 9 inches = 729 cubic inches

Notice we stated the volume in terms of cubic inches. In previous chapters when we measured for area, we stated our answers in terms of square inches or square feet or square yards or whatever measurement we were using at the time. When measuring distance, our labels were inches, feet, yards, or whatever measurement we were using at the time.

Distance tells us how far an object is, area tells us how much territory something covers, and volume tells us how much something can hold. For example, let's say we have a rectangular pool that's 15 feet wide, 40 feet long, and 10 feet deep. The distance around the pool, which is called its perimeter, is 2(15 feet) + 2(40 feet) = 110 feet. The amount of ground the pool covers is called its area; in this case it is (15 feet)(40 feet) = 600 square feet. The amount of space the pool takes up in the ground, which is the amount of space to be filled by water, is called its volume. In this case its volume is (15 feet)(40 feet)(10 feet) = 6,000 cubic feet. As you work your way through the rest of this chapter, please be careful to label your answers correctly.

Example 1:

Find the volume of a cube with an edge of $4\frac{1}{2}$ feet.

Solution:

Volume $= e^3 = (4.5 \text{ feet})^3 = (4.5 \text{ feet})(4.5 \text{ feet})(4.5 \text{ feet}) = 91.125$ cubic feet

Example 2:

Find the length of an edge of a cube whose volume is 4,096 cubic inches.

Solution:

This problem is the reverse of the last one. Here we're given the volume and are asked to find the measure of its edges.

$V = e^3$	Substitute 4,096 for the V.
$4,096 = e^3$	Take the cube root of each side of the equation.
$e = \sqrt[3]{4,096} = 16$ inches	

What is the surface area of the top of a cube that has an edge of 5 inches? The surface area measures the outside of the cube, not the inside. To find the surface area of any three-dimensional figure, we have to find the dimensions of all of its sides. Then we find the area of each side, and add all the areas together. Our question specifically asks only for the top, not all the sides.

The surface area is 25 square inches. OK, what is the entire surface area of this cube? (Remember, a cube has six equal sides.) The surface area of this cube is 25 square inches × 6 = 150 square inches.

Surface area of a cube formula
SA = $6e^2$

Example 3:

Find the surface area of a cube with an edge of $3\frac{1}{4}$ feet.

Solution:

Surface area = 6(*lw*) or $6e^2$ = $6(3.25)^2$ = 6(3.25 feet)(3.25 feet) = 6(10.5625 feet) = 63.375 square feet

Notice that we didn't round off the 10.5625 to fewer decimal places. Had we done so, our answer would have been different and not as accurate as the result we got. When we need to round an answer to a specific number of decimal places, we don't round until we get the final answer. If we round off numbers while we're still performing arithmetic operations, we create a greater rounding error, which makes our answer farther and farther from the exact answer. So control the urge to round off decimals until the end of the problem. We actually prefer to use fractions instead of decimals to get more accurate answers, but doing so makes the arithmetic more tedious. In the real world, most people prefer to let their calculators do the work for them, so we will perform most of our arithmetic operations using decimals.

Here's something a little different. If the surface area of a cube is 24 square inches, find the volume.

We'll work backward to find the measure of an edge of the cube and then plug it into the formula for the volume of a cube.

SA = $6e^2$ Substitute 24 for the surface area and solve for *e*.

24 = $6e^2$ Divide both sides of the equation by 6.

4 = e^2 Take the square root of both sides of the equation.

e = 2 inches

Volume = e^3 Substitute 2 for *e*.

V = (2 inches)^3

V = 8 cubic inches

Example 4:

Find the volume of a cube that has a surface area of 29.04 square yards.

Solution:

Surface area = $6e^2$ Substitute 29.04 for surface area.

29.04 = $6e^2$ Divide both sides of the equation by 6.

4.84 = e^2 Take the square root of both sides of the equation.

e = 2.2 yards

Volume = e^3 Substitute 2.2 for *e*.

V = 2.2^3

V = 10.648 cubic yards

Example 5:

The surface area of a cube is 486 square inches. Find the length of its edges.

Solution:

This is the same type of problem as the last one, except it's in reverse order.

$SA = 6e^2$ Substitute 486 for surface area and solve for *e*.

$486 = 6e^2$ Divide both sides of the equation by 6.

$81 = e^2$ Take the square root of both sides of the equation.

$e = \sqrt{81} = 9$

Rectangular Solids

A rectangular solid is a uniform solid whose base is a rectangle and whose height is perpendicular to its base.

Volume of a rectangular solid formula

Volume = (length)(width)(height) or $V = lwh$

Is a cube a rectangular solid?

     Yes! It's a rectangular solid that happens to have an identical length, width, and height. The following illustration is an example of a rectangular solid that does not have identical lengths, widths, and heights.

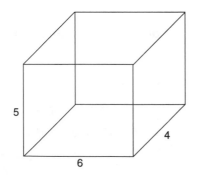

What is the volume of this box?

Volume = lwh = 4(6)(5) = 120

Example 6:

Find the volume of a box with these dimensions: length 12 feet, width 5 feet, and height 2 feet.

Solution:

Volume = lwh = 12 feet × 5 feet × 2 feet = 120 cubic feet

Example 7:

The volume of a rectangular solid is 150 cubic inches. The length is 10 inches, and the width is 5 inches. Find the height.

Solution:

We'll begin by using the given information and solve for the height.

V = lwh	Substitute 150 for V, 10 for l, and 5 for w.
150 = (10)(5)h	Multiply 10 times 5.
150 = 50h	Divide both sides of the equation by 50.
3 = h	The height is 3 inches.

Example 8:

The volume of a rectangular box is 216 cubic centimeters. The width is twice the height, and the length is twice the width. Find the dimensions.

Solution:

We don't know the dimensions of the length, width, or height, but we do know how they're related. The length is based on the width, which is based on the height, so we'll have to begin by letting x represent the height.

Let x = the measure of the height of the box.

Let $2x$ = the measure of the width of the box.

Let $2(2x) = 4x$ = the measure of the length of the box.

V = lwh	Substitute 216 for V, x for h, $2x$ for w, and $4x$ for l.
216 = ($4x$)($2x$)(x)	Remember to add the exponents when multiplying.
216 = $8x^3$	Divide both sides of the equation by 8.
27 = x^3	Take the cube root of both sides of the equation.
3 = x	

Let x = the measure of the height of the box. The height is 3 centimeters.

Let $2x$ = the measure of the width of the box. The width is 6 centimeters.

Let $2(2x) = 4x$ = the measure of the length of the box. The length is 12 centimeters.

Like the cube, other rectangular solids have six surfaces: a top and bottom, a front and back, and two sides. Obviously the top and the bottom of a rectangular solid have equal surface areas. Similarly, the front and back have equal surface areas, as do the two sides. See if you can sketch a rectangular solid with a length of 12 feet, a width of 5 feet, and a height of 2 feet. Then find the surface area of (a) the top and bottom; (b) front and back; and (c) the two sides.

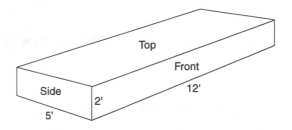

a. The surface area of the top = lw = (12 feet)(5 feet) = 60 square feet. The surface area of the bottom = 60 square feet.

b. The surface area of the front = lh = (12 feet)(2 feet) = 24 square feet. The surface area of the back = 24 square feet.

c. The surface area of a side = wh = (5 feet)(2 feet) = 10 square feet. The other side is also 10 square feet.

What is the surface area of the entire rectangular solid?

The total surface area = 2(the surface area of the top) + 2(the surface area of the front) + 2(the surface area of a side) = 2(60 square feet) + 2(24 square feet) + 2(10 square feet) = 120 square feet + 48 square feet + 20 square feet = 188 square feet.

$SA = 2lw + 2lh + 2wh$

Example 9:

Find the surface area of a rectangular solid with length 10 inches, width 4 inches, and height 7 inches.

Solution:

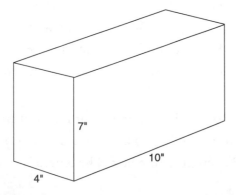

The surface area of the top = lw = (10 inches)(4 inches) = 40 square inches.

The surface area of the front = lh = (10 inches)(7 inches) = 70 square inches.

The surface area of the side = wh = (4 inches)(7 inches) = 28 square inches.

The total surface area of the rectangular solid = 2(surface area of the top) + 2(surface area of the front) + 2(the surface area of a side) = 2(40 square inches) + 2(70 square inches) + 2(28 square inches) = 80 square inches + 140 square inches + 56 square inches = 276 square inches.

Example 10:

The surface area of a rectangular solid is 388 square feet. The length is 20 feet, the width is 7 feet. Find its height.

Solution:

We'll substitute all the given information into the formula for the surface area of a rectangular solid and solve for h.

$SA = 2lw + 2lh + 2wh$ Substitute 388 for SA, 20 for l, and 7 for w.

$388 = 2(20)(7) + 2(20)h + 2(7)h$ Multiply.

$388 = 280 + 40h + 14h$ Combine like terms: $40h + 14h = 54h$.

$388 = 280 + 54h$ Subtract 280 from both sides of the equation.

$108 = 54h$ Divide both sides of the equation by 54.

$2 = h$ The height is 2 feet.

SELF-TEST 1

1. What is the volume of a cube with an edge of

    a. 5 feet?   b. 7.5 inches?   c. 1.4 yards?

2. Find the surface area of a cube with an edge of

    a. 2 feet   b. 5.5 yards   c. 19.6 inches

3. Find the volume of a cube that has a surface area of

    a. 150 square inches   b. 6 square yards   c. 65.34 square inches

4. What is the volume of the following box?

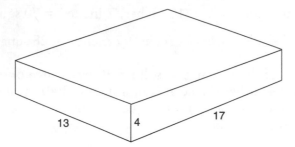

5. What is the volume of a rectangular solid with a length of 17 feet, a width of 12 feet, and a height of 10 feet?

6. Find the surface area of a rectangular solid with

   a. a length of 6 inches, a width of 4 inches, and a height of 2 inches

   b. a length of 18 feet, a width of 9 feet, and a height of 6 feet

ANSWERS

1. a. Volume = e^3 = 5^3 = 125 cubic feet

   b. Volume = e^3 = $(7.5)^3$ = 421.875 cubic inches

   c. Volume = e^3 = $(1.4)^3$ = 2.744 cubic yards

2. a. Surface area = $6e^2$ = $6(2)^2$ = 6(4) = 24 square feet

   b. Surface area = $6e^2$ = $6(5.5)^2$ = 6(30.25) square yards = 181.5 square yards

   c. Surface area = $6e^2$ = $6(19.6)^2$ = 6(384.16) square inches = 2,304.96 square inches

3. a. Surface area = $6e^2$ Substitute 150 for the surface area.

   $150 = 6e^2$ Divide both sides of the equation by 6.

   $25 = e^2$ Take the square root of both sides of the equation.

   $e = 5$ inches

   Volume = e^3 = 5^3 = 125 cubic inches

   b. The surface area = $6e^2$ Substitute 6 for the surface area.

   $6 = 6e^2$ Divide both sides of the equation by 6.

   $1 = e^2$ Take the square root of both sides of the equation.

   $e = 1$ yard

   $V = e^3 = 1^3 = 1$ cubic yard

c. Surface area = $6e^2$ Substitute 65.34 for the surface area.

$65.34 = 6e^2$ Divide both sides of the equation by 6.

$10.89 = e^2$ Take the square root of both sides of the equation.

$e = 3.3$ inches

$V = e^3 = (3.3)^3 = 35.937$ cubic inches

4. Volume = lwh = (17)(13)(4) = 884

5. Volume = lwh = (17 feet)(12 feet)(10 feet) = 2,040 cubic feet

6. We'll begin by finding the surface area of the top and double it to include the bottom. Then we'll find the surface area of the front and double that to include the back. Finally we'll find the surface area of a side and double that to include the other side. Once we have all this information, we'll add these amounts to find the surface area of the entire rectangular solid.

   a. Surface area of top = lw = (6 inches)(4 inches) = 24 square inches

   Surface area of front = lh = (6 inches)(2 inches) = 12 square inches

   Surface area of side = wh = (4 inches)(2 inches) = 8 square inches

   Surface area = 2(24) + 2(12) + 2(8) = 88 square inches

   b. Surface area of top = lw = (18 feet)(9 feet) = 162 square feet

   Surface area of front = lh = (18 feet)(6 feet) = 108 square feet

   Surface area of side = wh = (9 feet)(6 feet) = 54 square feet

   Surface area = 2(162 square feet) + 2(108 square feet) + 2(54 square feet) = 324 square feet + 216 square feet + 108 square feet = 648 square feet

Pyramids

The world's most famous pyramids are, of course, in Egypt. Even though we may be thousands of miles away, if we know some of their dimensions, we can still calculate their impressive volume and surface area.

First a general definition: *A pyramid is a geometric solid having any polygon as one face, where all the other faces are triangles meeting at a common vertex. The pyramid is named after the polygon forming the face from which the triangles start.*

To keep things as simple as possible, the face of all the pyramids we'll consider will be its base. In addition, we'll work with pyramids having just three bases: the triangle-based pyramid (tetrahedron), the rectangle-based pyramid, and the right square–based pyramid.

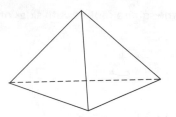

Volume of a pyramid formula is

$$V = \frac{1}{3}Bh$$

where B is the area of the base and h is the height of the pyramid.
Find the volume of a pyramid with a base area of 10 and a height of 15.

$$\text{Volume} = \frac{1}{3}Bh = \frac{1}{3}(10)(15) = 50$$

Example 11:

Find the volume of the following pyramid.

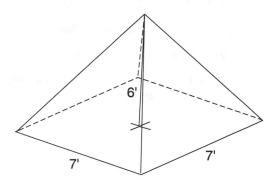

Solution:

This pyramid has a base that's a square. We know this because the length and width are equal. B is the area of the base, which is $(7)(7)$. The height is 6. The volume of this pyramid is $V = \frac{1}{3}Bh = \frac{1}{3}(7)(7)(6) = 98$ cubic feet.

Example 12:

The volume of a pyramid with a square base is 48 cubic yards. The height is 4 yards. What are the dimensions of its base?

Solution:

To find the length of the sides of the base, we have to find the value of B, the area of the base. To find B, we have to use the formula for the area of a square: $A = s^2$. In this case we'll say $B = s^2$. We'll begin by substituting the given information into the formula for the volume of a pyramid and solve for B.

$V = \frac{1}{3}Bh$	Substitute 48 for V and 4 for h.
$48 = \frac{1}{3}B(4)$	Multiply both sides of the equation by 3 to clear out the fraction.
$144 = 4B$	Divide both sides of the equation by 4.
$36 = B$	The area of the base is 36 square yards.

The dimensions of the base are 6 yards by 6 yards.

Example 13:

The volume of a pyramid with a square base is 48 feet. The pyramid is 9 feet tall. Find the length of the sides of its base.

Solution:

We're given the volume and the height, so we'll start by substituting the given values into the formula for the volume of a pyramid.

$V = \frac{1}{3}Bh$	Substitute 48 for V and 9 for h.
$48 = \frac{1}{3}B(9)$	Multiply 9 times $\frac{1}{3}$.
$48 = 3B$	Divide both sides of the equation by 3.
$B = 16$	
$B = s^2$	Substitute 16 for B and solve for s.
$16 = s^2$	Take the square root of both sides of the equation.
$s = 4$ feet	

Now we'll find the surface area of pyramids. To do that, we need to know three things: the area of the base of one side, the height of that side, and the number of sides. Then we plug those numbers into this formula:

Surface area of a pyramid formula

$\frac{1}{2}$(area of the base)(height)(number of sides) or $SA = \frac{1}{2}Bhs$

Example 14:

The area of the base of a four-sided pyramid is 12 and its height is 16. Find the surface area of this pyramid.

Solution:

$$SA = \frac{1}{2}(12)(16)(4) = 384$$

Example 15:

The surface area of a three-sided pyramid is 1,125 square feet. Find the height of the pyramid if the base of the pyramid has sides 5 feet in length.

Solution:

Before we can use the formula for the surface area of a pyramid, we have to find *B*, the area of the base of a pyramid. $B = s^2$. We already know $s = 5$, so $B = 5^2$ or 25 square feet.

$SA = \frac{1}{2}Bhs$	Substitute 1,125 for SA, 25 for *B*, and 3 for *s*.
$1{,}125 = \frac{1}{2}(25)(h)(3)$	Multiply both sides of the equation by 2.
$2{,}250 = 75h$	Divide both sides of the equation by 75.
$h = 30$	The height of the pyramid is 30 feet.

SELF-TEST 2

1. Find the volume of each of these pyramids:
   a. base area of 15 square inches and a height of 9 inches
   b. base area of 10 square feet and a height of 7 feet
   c. base area of 42 square yards and a height of 10 yards

2. Find the volume of the following pyramid.

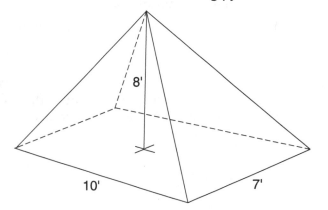

3. One side of a six-sided pyramid has a base with an area of 9 square feet and a height of 7 feet. Find the surface area of this pyramid.

4. One side of a three-sided pyramid has a base with an area of 34 square inches and a height of 12 inches. Find the surface area of this pyramid.

ANSWERS

1. a. Volume $= \frac{1}{3} Bh$ Substitute 15 for B and 9 for h.

   $V = \frac{1}{3}(15)(9) = 45$ cubic inches

   b. Volume $= \frac{1}{3} Bh$ Substitute 10 for B and 7 for h.

   $V = \frac{1}{3}(10)(7) = \frac{70}{3} = 23\frac{1}{3}$ cubic feet

   c. Volume $= \frac{1}{3} Bh$ Substitute 42 for B and 10 for h.

   $V = \frac{1}{3}(42)(10) = 140$ cubic yards

2. Volume $= \frac{1}{3} Bh$ $B = (10)(7)$, $h = 8$

   $V = \frac{1}{3}(10)(7)(8) = \frac{560}{3} = 186\frac{2}{3}$ cubic feet

3. Surface area $= \frac{1}{2} Bhs$ $n = 6, B = 9, h = 7, s = 6$

   $SA = \frac{1}{2}(9)(7)(6) = 189$ square feet

4. Surface area $= \frac{1}{2} Bhs$ $n = 3, B = 34, h = 12$

   $SA = \frac{1}{2}(34)(12)(3) = 612$ square inches

Spheres

Spheres are part of our everyday lives, especially if we play ball. Basketballs, tennis balls, baseballs, and golf balls are all spheres. Earth, the Moon, and the other planets and moons of our solar system are also spheres, although not perfect spheres. *A sphere is either the shape of a surface in three dimensions that is everywhere the same distance from a single fixed point or the solid shape enclosed by that surface.*

In this section, we'll also work with half spheres, known as hemispheres. The following figure shows both.

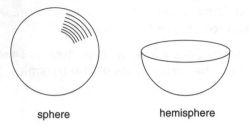

sphere   hemisphere

You know how things work around here. First we'll ask you to find the volume and then the surface area.

Volume of a sphere formula is

$$V = \frac{4}{3}\pi r^3$$

What would be the formula for the volume of a hemisphere? Remember, a hemisphere is half a sphere.

Volume of a hemisphere formula

$$V = \left(\frac{1}{2}\right)\left(\frac{4}{3}\right)\pi r^3, \text{ which is } \frac{2}{3}\pi r^3, \text{ so } V = \frac{2}{3}\pi r^3$$

Example 16:
Find the volume of a hemisphere with a radius of 10 inches.

Solution:

$$V = \frac{2}{3}\pi r^3 = (0.67)(3.14)(10)^3 = (2.1038)(1,000) = 2,103.8 \text{ cubic inches}$$

Example 17:
Find the volume of a sphere with a diameter of 30 inches.

Solution:

The formula for the volume of a sphere uses the radius, not the diameter, of the sphere. We'll find the radius by dividing the diameter by 2.

$$r = \frac{d}{2} = \frac{30}{2} = 15$$

$$\text{Volume} = \frac{4}{3}\pi r^3 = (1.33)(3.14)(15)^3 = (4.1762)(3,375) = 14,094.7 \text{ cubic inches}$$

Example 18:

Find the volume of a hemisphere with a radius of 4 feet.

Solution:

$$V = \frac{2}{3}\pi r^3 = (.67)(3.14)(4)^3 = (2.1038)(64) = 134.643 \text{ cubic feet}$$

We've looked at the surface area of other three-dimensional solids, so why not a sphere?

Surface area of a sphere formula
SA = $4\pi r^2$

Example 19:

Find the surface area of a sphere with a radius of 5 yards.

Solution:

SA = $4\pi r^2$ = $4(3.14)(5)^2$ = $4(3.14)(25)$ = 314 square yards

Example 20:

Find the surface area of a sphere with a diameter of 12.2 yards.

Solution:

First we have to find the radius of the sphere.

$$r = \frac{d}{2} = \frac{12.2}{2} = 6.1$$

SA = $4\pi r^2$ = $4(3.14)(6.1)^2$ = $(12.56)(37.21)$ = 467.358 square yards

Can you figure out the formula for the surface area of a hemisphere? (Don't worry about the flat part; just get the spherical part.)

Surface area of a hemisphere formula

SA = $\left(\frac{1}{2}\right)(4\pi r^2)$ = $2\pi r^2$, or SA = $2\pi r^2$

Example 21:

Find the surface area of a hemisphere with a radius of 2.5 feet.

Solution:

SA = $2\pi r^2$ = $2(3.14)(2.5)^2$ = $(6.28)(6.25)$ = 39.25 square feet

SELF-TEST 3

1. Find the volume for a sphere with a radius of

   a. 2 feet   b. 9 inches   c. 3.5 yards

2. Find the volume for a sphere with a diameter of 15 inches.

3. Find the volume of a hemisphere with a radius of 18 feet.

4. Find the surface area of a sphere with a radius of

   a. 30 feet   b. 8.5 feet   c. 14 meters

5. Find the surface area of a hemisphere with a radius of

   a. 17 feet   b. 2.9 inches

ANSWERS

1. a. $V = \frac{4}{3}\pi r^3 = (1.33)(3.14)(2)^3 = 4.1762(8) = 33.4096$ cubic feet

   b. $V = \frac{4}{3}\pi r^3 = (1.33)(3.14)(9)^3 = 4.1762(729) = 3{,}044.45$ cubic inches

   c. $V = \frac{4}{3}\pi r^3 = (1.33)(3.14)(3.5)^3 = 4.1762(42.875) = 179.055$ cubic yards

2. This formula calls for the radius, not the diameter. First find the radius by dividing the diameter by 2.

   $r = \dfrac{d}{2} = \dfrac{15}{2} = 7.5$

   $V = \frac{4}{3}\pi r^3 = (1.33)(3.14)(7.5)^3 = 4.1762(421.875) = 1{,}761.83$ cubic inches

3. $V = \frac{2}{3}\pi r^3 = (0.67)(3.14)(18)^3 = (2.1038)(5{,}832) = 12{,}269.4$ cubic feet

4. a. $SA = 4\pi r^2 = 4(3.14)(30)^2 = (12.56)(900) = 11{,}304$ square feet

   b. $SA = 4\pi r^2 = 4(3.14)(8.5)^2 = (12.56)(72.25) = 907.46$ square inches

   c. $SA = 4\pi r^2 = 4(3.14)(14)^2 = (12.56)(196) = 2{,}461.76$ square meters

5. a. $SA = 2\pi r^2 = 2(3.14)(17)^2 = (6.28)(289) = 1{,}814.92$ square feet

   b. $SA = 2\pi r^2 = 2(3.14)(2.9)^2 = (6.28)(8.41) = 52.81$ square inches

Cylinders

Cylinders are familiar objects, ranging from automobile motor parts (generally four to eight) to toilet paper holders, rolling pins, and baseball bats. If you want to get picky, these last three objects are not purely cylindrical in shape, but they come close enough. *A cylinder is a uniform solid whose base is a circle.*

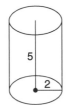

The cylinder shown above has a radius of 2 and a height of 5.

Volume of a cylinder formula
$V = \pi r^2 h$

where r is the radius and h is the height of the cylinder

See if you can find the volume of this cylinder.

$V = \pi r^2 h = (3.14)(2)^2(5) = 62.8$

Example 22:
Find the volume of a cylinder with a radius of 3.8 feet and a height of 4.6 feet.

Solution:
Volume $= \pi r^2 h = (3.14)(3.8)^2(4.6) = (3.14)(14.44)(4.6) = 208.571$ cubic feet

Example 23:
Find the volume of a cylinder with a diameter of 10 inches and a height of 29 inches.

Solution:
First we'll begin by finding the radius.
$$r = \frac{d}{2} = \frac{10}{2} = 5$$
Volume $= \pi r^2 h = (3.14)(5)^2(29) = (3.14)(25)(29) = 2,276.5$ cubic inches

Example 24:

Find the surface area of a cylinder with a radius of 12 inches and a height of 5 inches.

Solution:

$SA = 2\pi rh = 2(3.14)(12)(5) = 6.28(60) = 376.8$ square inches

Example 25:

Find the surface area of a cylinder with a diameter of 9 feet and a height of 13.4 feet.

Solution:

Start by finding the radius.

$$r = \frac{d}{2} = \frac{9}{2} = 4.5$$

$SA = 2\pi rh = 2(3.14)(4.5)(13.4) = 6.28(60.3) = 378.684$ square feet

SELF-TEST 4

1. Find the volume of a cylinder with a

   a. radius of 12 feet and a height of 9 feet.

   b. radius of 4.5 yards and a height of 12 yards.

   c. diameter of 16 inches and a height of 23 inches.

2. Find the surface area of a cylinder with the same dimensions as those in problem 1.

ANSWERS

1. a. Volume $= \pi r^2 h = (3.14)(12)^2(9) = (3.14)(144)(9) = 4,069.44$ cubic feet

   b. Volume $= \pi r^2 h = (3.14)(4.5)^2(12) = (3.14)(20.25)(12) = 763.02$ cubic yards

   c. Volume $= \pi r^2 h = (3.14)(8)^2(23) = (3.14)(64)(23) = 4,622.08$ cubic inches

2. a. SA $= 2\pi rh = 2(3.14)(12)(9) = (6.28)(108) = 678.24$ square feet

   b. SA $= 2\pi rh = 2(3.14)(4.5)(12) = (6.28)(54) = 339.12$ square yards

   c. SA $= 2\pi rh = 2(3.14)(8)(23) = (6.28)(184) = 1,155.52$ cubic inches

Cones

You know cones. We eat ice cream cones and snow cones, and we see orange traffic cones whenever there is highway construction. And we've all seen conical headgear ranging from party hats to coneheads and dunce caps.

A cone is the three-dimensional shape formed by a straight line when one end is moved around a simple closed curve, while the other end of the line is kept fixed at a point that is not in the plane of the curve. The following figure is a cone with a radius of 3 and a height of 6.

Volume of a cone formula

$$V = \frac{1}{3}\pi r^2 h$$

See if you can find the volume of this cone.

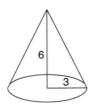

Solution:

$$\text{Volume} = \frac{1}{3}\pi r^2 h = (.33)(3.14)(3)^2(6) = (1.0362)(54) = 55.95$$

Example 26:

Find the volume of a cone with a diameter of 10 inches and a height of 27 inches.

Solution:

$$\text{Volume} = \frac{1}{3}\pi r^2 h = (.33)(3.14)(5)^2(27) = (1.0362)(675) = 699.435 \text{ cubic inches}$$

SELF-TEST 5

1. Find the volume of a cone with a radius of 12 inches and a height of 25 inches.

2. Find the volume of a cone with a radius of 3.5 feet and a height of 6 feet.

3. Find the volume of a cone with a diameter of 8 yards and a height of 13 yards.

1. Volume $= \frac{1}{3}\pi r^2 h = (.33)(3.14)(12)^2(25) = (1.0362)(144)(25) = 3{,}730.32$ cubic inches

2. Volume $= \frac{1}{3}\pi r^2 h = (.33)(3.14)(3.5)^2(6) = (1.0362)(12.25)(6) = 76.1607$ cubic feet

3. Volume $= \frac{1}{3}\pi r^2 h = (.33)(3.14)(4)^2(13) = (1.0362)(16)(13) = 215.53$ cubic yards

Complex Geometric Solids

So far we've found the volume and surface area of cubes, rectangular solids, pyramids, spheres, cylinders, and cones. Now we get to mix and match, so to speak. Let's start out by finding the volume of a cube atop a rectangular solid. We want to find the volume of the entire complex figure.

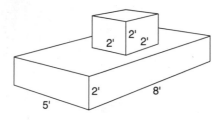

Solution:

To find the volume of this entire geometric solid, we have to find the volume of the cube and the volume of the rectangular solid, and add the two together.

Volume of the cube $= e^3 = (2)^3 = 8$ cubic feet

Volume of the rectangular solid $= lwh = (8 \text{ feet})(5 \text{ feet})(2 \text{ feet}) = 80$ cubic feet

Volume of the complex figure $= 8$ cubic feet $+ 80$ cubic feet $= 88$ cubic feet

Finding the surface area of this complex figure is a lot more complicated than finding the volume, but let's give it a shot anyway. We have to watch out for two things. First, let's not consider the bottom, or underside, of the rectangular solid as part of the surface area of the figure. Also, we'll have to be careful not to count the surface area of the bottom of the square, which rests on top of the rectangular solid, as part of the "exposed" surface area of the figure.

Let's find the (exposed) surface area of the cube. Then we'll do the same for the rectangular solid, and add the two surface areas together.

Exposed surface area of the cube $= 5e^2 = 5(2)^2 = 5(4) = 20$ square feet.

Notice we didn't use SA = $6e^2$ because we're not counting the bottom of the box as part of the surface area.

The surface area of the rectangular solid:

Surface area of the top = lw = (8)(5) = 40 square feet

We have to subtract the 4 square feet occupied by the bottom of the cube:

40 square feet − 4 square feet = 36 square feet

Surface area of the front = lh = (8)(2) = 16 square feet

Surface area of the back = 16 square feet

Surface area of the side = wh = (5)(2) = 10 square feet

There are two sides, so the surface area of the sides is 2(10) = 20 square feet.

Surface area of the rectangular solid = 36 square feet + 16 square feet + 16 square feet + 20 square feet = 88 square feet

Surface area of the complex solid = 88 square feet + 20 square feet = 108 square feet

Example 27:

Here's one that calls for a little imagination. A cylindrical shaft 3 feet in diameter is cut through a sphere with a diameter of 20 feet. Find the volume of the sphere (just the solid part, without the shaft).

Solution:

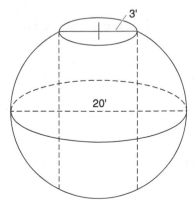

Here's the plan. Calculate the volume of the sphere. Then calculate the volume of the cylindrical shaft and subtract it from the volume of the sphere.

Volume of the sphere = $\frac{4}{3}\pi r^3$ = (1.33)(3.14)(10)^3 = (4.1762)(1,000) = 4,176.2 cubic feet

Volume of the cylinder = $\pi r^2 h$ = (3.14)(1.5)^2(20) = (3.14)(2.25)(20) = 141.3 cubic feet

Volume of the sphere without the shaft = 4,176.2 cubic feet − 141.3 cubic feet = 4,034.9 cubic feet

Example 28:

Find the volume of the following complex figure consisting of a pyramid atop a cube. The pyramid has a height of 10 inches, and the area of its base is 28 square inches. The cube has an edge of 18 inches. (Hint: notice the figure doesn't have a bottom.)

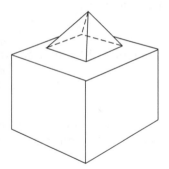

Solution:

We want to find the volume of the pyramid, the volume of the cube, and then add the two together.

Volume of the pyramid = $\frac{1}{3}Bh = \frac{1}{3}(28)(10) = \frac{280}{3} = 93\frac{1}{3}$ cubic inches

Volume of the cube = e^3 = (18)^3 = 5,832 cubic inches

Volume of the complex figure = $93\frac{1}{3}$ cubic inches + 5,832 cubic inches = $5,925\frac{1}{3}$ cubic inches

SELF-TEST 6

1. A cube with an edge of 2 feet protrudes from the side of a cube with an edge of 5 feet. Assume the cubes don't have a bottom.

   a. Find the volume of this geometric solid.

   b. Find the surface area of this geometric solid.

2. A hemisphere with a diameter of 15 inches sits on a cube (diameter side down) with an edge of 15 inches. Assume the hemisphere doesn't have a bottom.

   a. Find the volume of this geometric solid.

   b. Find the surface area of this geometric solid.

3. Find the volume of the following geometric shape.

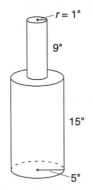

ANSWERS

1. a. First we'll find the volume of the smaller cube, then the volume of the larger cube, and finally, we'll add their volumes to obtain the volume of the geometric solid.

   Volume of the smaller cube = $e^3 = (2)^3 = 8$ cubic feet

   Volume of the larger cube = $e^3 = (5)^3 = 125$ cubic feet

   Volume of the geometric solid = 8 cubic feet + 125 cubic feet = 133 cubic feet

   b. Let's look first at the surface area of the smaller cube. Because one of its six surfaces lies directly against the larger cube, it has, in effect, only five surface areas.

   Surface area of the smaller cube = $5e^2 = 5(2)^2 = 5(4) = 20$ square feet

   We can eliminate the bottom surface area of the larger cube because presumably it rests on the ground. Also, we'll eliminate the part of the surface area on one side that it shares with one side of the smaller cube, an area of 4 square feet.

   Surface area of the larger cube = $5e^2 - 4 = 5(5)^2 - 4 = 5(25) - 4 = 125 - 4 = 121$ square feet

   Surface area of the solid geometric shape = 20 square feet + 121 square feet = 141 square feet

2. a. Volume of the hemisphere $= \frac{2}{3}\pi r^3 = (.67)(3.14)(7.5)^3 = (2.1038)(421.875) = 887.541$ cubic inches

   Volume of the cube $= e^3 = (15)^3 = 3,375$ cubic inches

   Volume of the geometric solid $= 887.541$ cubic inches $+ 3,375$ cubic inches $= 4,262.54$ cubic inches

   b. Surface area of the hemisphere $= 2\pi r^2 = 2(3.14)(7.5)^2 = (6.28)(56.25) = 353.25$ square inches

   To find the surface area of the cube, we'll need to eliminate the bottom surface area as well as the surface area on the top that is occupied by the hemisphere.

   Hemisphere area $= \pi r^2 = (3.14)(7.5)^2 = (3.14)(56.25) = 176.625$ square inches

   Surface area of the cube without the bottom $= 5e^2 = 5(15)^2 = (5)(225) = 1,125$ square inches

   Surface area of the cube with an opening for the hemisphere $= 1,125$ square inches $-$ 176.63 square inches $= 948.37$ square inches

   Surface area of the complex figure $= 353.25$ square inches $+ 948.37$ square inches $=$ $1,301.62$ square inches

3. First we'll find the volume of the larger cylinder and then the volume of the smaller cylinder.

   Volume of the larger cylinder $= \pi r^2 h = (3.14)(5)^2(15) = (3.14)(25)(15) = (3.14)(375) =$ $1,177.5$ cubic inches

   Volume of the smaller cylinder $= \pi r^2 h = (3.14)(1)^2(9) = (3.14)(9) = 28.26$ cubic inches

   Volume of the geometric shape $= 1,177.5$ cubic inches $+ 28.26$ cubic inches $= 1,205.76$ cubic inches

Applications

Let's see if we can put some of this knowledge about solid geometry to practical use. Each of these problems involves finding the surface area and volume of a geometric shape.

Example 29:

How many gallons of whitewash would you need to paint a cylindrical water tank with a radius of 80 feet and a height of 200 feet, if 1 gallon is needed for every 500 square feet?

Solution:

Surface area = $2\pi rh$ = 2(3.14)(80 feet)(200 feet) = 6.28(16,000 square feet) = 100,480 square feet

One gallon covers 500 square feet, so we'll divide 100,480 square feet by 500 to find the number of gallons needed to cover the water tank.

Number of gallons needed = (100,480) ÷ (500) = 200.96 gallons

Example 30:

A cube of marble with an edge of 10 feet contains a hollow sphere 6 feet in diameter. (In other words, there's a spherical hole inside a marble cube.) How much does the cube weigh if a cubic foot of marble weighs 38 pounds?

Solution:

First find the volume of a solid marble cube, then find the volume of the sphere, and subtract it from the volume of the marble. Finally, multiply that volume by 38.

Volume of the cube = e^3 = $(10)^3$ = 1,000 cubic feet

Radius of the sphere = 6 ÷ 2 = 3

Volume of the sphere = $\frac{4}{3}\pi r^3$ = (1.33)(3.14)(3)^3 = (1.33)(3.14)(27) = (4.1762)(27) = 112.757 cubic feet

Volume of the hollowed cube = 1,000 cubic feet − 112.757 cubic feet = 887.243 cubic feet

Weight of the hollowed cube = 887.243 × 38 pounds = 33,715.2 pounds

Example 31:

If hydrochloric acid is being poured into a cylindrical container at the rate of 10 cubic meters per minute, how many minutes would it take to fill up a tank that is 19 meters in diameter and has a height of 23 meters?

Solution:

First find the volume of the tank (in cubic meters), then divide that number by 10 cubic meters.

Volume = $\pi r^2 h$ = (3.14)(9.5)^2(23) = 3.14(90.25)(23) = (3.14)(2,075.75) = 6,517.86 cubic meters

Number of minutes to fill up the tank = (6,517.86 cubic meters) ÷ (10) = 651.786 minutes, which is just under 11 hours

Example 32:

Let's see how much ice cream there actually is in an ice cream cone. The ice cream cone in the following illustration consists of a cone topped by a hemisphere. See if you can figure out how many cubic inches of ice cream you actually get.

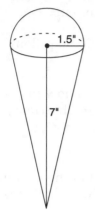

Solution:

We have to add the volume of the cone to the volume of the hemisphere.

Volume of the cone $= \dfrac{1}{3}\pi r^2 h = (.33)(3.14)(1.5)^2(7) = (1.0362)(2.25)(7) = 16.3202$ cubic inches

Volume of the hemisphere $= \dfrac{2}{3}\pi r^3 = (.67)(3.14)(1.5)^3 = (2.1038)(3.375) = 7.10033$ cubic inches

Amount of ice cream $= 16.3202 + 7.10033 = 23.4205$ cubic inches

Example 33:

A sculptor is pouring plastic into a hollow mold of a pyramid with a base area of 30 square feet and a height of 10 feet. If the plastic costs $18 per cubic foot, how much will it cost to fill the mold?

Solution:

Volume $= \dfrac{1}{3}Bh = \dfrac{1}{3}(30)(10) = \dfrac{1}{3}(300) = 100$ cubic feet

100 cubic feet $\times$ \$18 $=$ \$1,800

SELF-TEST 7

1. How large a tarpaulin would be needed (in square feet) to cover a cube with an edge of 17 feet? Assume the bottom of the cube is not covered.

2. Concrete is being poured for the foundation of a building that is 60 feet long and 28 feet wide. If the foundation will be 3 feet deep, how much will it cost if the price of concrete is $3 per cubic foot?

3. a. How much water can be held by a cylindrical tank that has a diameter of 90 feet and a height of 120 feet?

   b. If it takes 1 gallon of paint to cover 750 square feet of surface area, how many gallons of paint will it take to cover the entire water tank?

4. A sculpture consists of a cylinder sitting on top of a cube. If the cube has an edge of 4 feet and the cylinder has a radius of 1 foot and a height of 5 feet, what is the total volume of the sculpture?

5. If 3 cubic feet of water a minute is being poured into a spherical fish tank with a diameter of 9 feet, how long will it take to fill up the tank?

6. You need to find the weight of a sphere with a cylindrical shaft drilled through it. The sphere has a diameter of 10 feet and the hole has a diameter of 4 feet. If the sphere weighs 40 pounds per cubic foot, find the weight of the sphere.

ANSWERS

1. Surface area $= 5e^2 = 5(17)^2 = 5(289) = 1{,}445$ square feet

2. Volume $= lwh = (60 \text{ feet})(28 \text{ feet})(3 \text{ feet}) = 5{,}040$ cubic feet

   5,040 cubic feet $\times$ $3 = $15,120$

3. a. Volume $= \pi r^2 h = (3.14)(45)^2(120) = (3.14)(2{,}025)(120) = 763{,}020$ cubic feet

   b. Surface area $= 2\pi rh = 2(3.14)(45)(120) = 6.28(5{,}400) = 33{,}912$ square feet

      Now that we know the number of square feet of surface area, we have to divide by 750 to find the number of gallons needed to cover the tower.

      $33{,}912 \div 750 = 45.216$ gallons

4. Volume of the cube $= e^3 = (4)^3 = 64$ cubic feet

   Volume of the cylinder $= \pi r^2 h = (3.14)(1)^2(5) = (3.14)(5) = 15.7$ cubic feet

   Volume of the sculpture $= 64$ cubic feet $+ 15.7$ cubic feet $= 79.7$ cubic feet

5. Volume $\frac{4}{3}\pi r^3 = (1.33)(3.14)(4.5)^3 = (4.1762)(91.125) = 380.556$ cubic feet

   $(380.556) \div (3) = 126.852$ minutes, or about 2 hours 7 minutes

6. First we'll find the volume of the solid sphere, then the volume of the cylindrical hole, which we'll subtract from the volume of the sphere.

   Volume of the solid sphere $= \frac{4}{3}\pi r^3 = (1.33)(3.14)(5)^3 = (4.1762)(125) = 522.025$ cubic feet

   Volume of the cylindrical shaft $= \pi r^2 h = (3.14)(2)^2(10) = (3.14)(40) = 125.6$ cubic feet

   Volume of the sphere with the shaft = 522.025 cubic feet − 125.6 cubic feet = 396.425 cubic feet

   40 pounds × 396.425 cubic feet = 15,857 pounds

6 Conic Sections

You may never have even heard of conic sections, but the chances are you know quite a lot about fast foods. If you work in McDonald's, Burger King, or Wendy's, the question customers are often asked is "Do you want fries with that?" At any pizzeria, the question most often asked is "How many slices?" But when you go to an ice cream parlor, no one asks how you want that cone sliced. Well, guess what? In this chapter we're going to be slicing up cones to form circles, ellipses, hyperbolas, and parabolas. When you have completed this chapter, you will be able to

- write the equations of circles, ellipses, hyperbolas, and parabolas

- determine the center, radius, and symmetry of circles

- find the coordinates of the focus and the equation of the directrix of parabolas

- find the coordinates of the vertices and the foci of hyperbolas

- work on applications involving conics

Before we get started, please try the following pretest. If you get all the problems correct, you can skip this chapter. Good luck!

PRETEST

1. Determine the center and the radius of $(x - 3)^2 + (y + 4)^2 = 49$. What type of conic section is this?

2. Find the equation of the circle with center (4,–1), and radius 7. Write your answer in standard form and general form.

3. Determine the center and radius of the circle $3x^2 + 3y^2 + 6x = 1$.

4. Find the equation of the parabola with focus (–3,–2) and directrix $x = 1$.

5. Determine the coordinates of the focus and the equation of the directrix of the given parabola. Graph the parabola.

   $x^2 = 8y$

6. Find the vertex, focus, and directrix of this parabola.

   $y^2 - 4y - 4x = 0$

7. Find the vertex, focus, and directrix of the parabola given by the following equation. Sketch its graph.

   $x^2 + 8x - y + 6 = 0$

8. Find the vertices and foci of the following ellipse.

   $\dfrac{x^2}{25} + \dfrac{y^2}{49} = 1$

9. Find the vertices and foci of the following ellipse.

   $4x^2 + y^2 - 8x + 4y - 8 = 0$

10. Find an equation for a hyperbola with foci $(3, 2 + \sqrt{5})$ and $(3, 2 - \sqrt{5})$ with vertices (3,0), and (3,4); asymptotes $y = \dfrac{2}{3}x$, $y = 4 - \dfrac{2}{3}x$.

11. Find the center, vertices, and foci of a hyperbola with the following equation. Graph the hyperbola.

    $x^2 - 9y^2 + 36y - 72 = 0$

12. A seal-beam headlight is in the shape of a paraboloid of revolution. The bulb, which is placed at the focus, is 1 inch from the vertex. If the depth is to be 2 inches, what is the diameter of the headlight at its opening?

13. A semielliptical arch over a tunnel for a road through a mountain has a major axis of 100 feet and a height at the center of 30 feet.

    a. Create a sketch for solving the problem. Draw a rectangular coordinate system on the tunnel with the center of the road entering the tunnel at the origin. Identify the coordinates of the known points.

    b. Find the equation of the elliptical tunnel.

    c. Determine the height of the arch 5 feet from the edge of the tunnel.

14. A cable TV receiving dish is in the shape of a paraboloid of revolution, as shown in the following figure. Find the location of the receiver, which is placed at the focus, if the dish is 10 feet across at its opening and 3 feet deep.

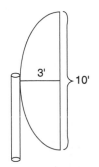

15. An arch in the form of half an ellipse is 40 feet wide and 15 feet high at the center. Find the height of the arch at intervals of 10 feet along its width.

ANSWERS

1. This is a circle with center: (3,–4), radius: 7.

2. $(x - 4)^2 + (y + 1)^2 = 7^2$ standard form

   $x^2 - 8x + 16 + y^2 + 2y + 1 = 49$

   $x^2 + y^2 - 8x + 2y - 32 = 0$ general form

3. $3x^2 + 3y^2 + 6x = 1$

   $x^2 + y^2 + 2x = \dfrac{1}{3}$

   $(x^2 + 2x) + y^2 = \dfrac{1}{3}$

   $(x^2 + 2x + 1) + y^2 = \dfrac{1}{3} + 1$

   $(x + 1)^2 + y^2 = \dfrac{4}{3}$

   center: (–1,0), radius: $\sqrt{\dfrac{4}{3}} = \dfrac{2}{\sqrt{3}} = \dfrac{2\sqrt{3}}{3}$

4. $F(-3,-2)$ directrix: $x = 1$ directrix $= h - p$, $h - p = 1$

   $p = -2$ The vertex is halfway between the focus and the directrix. V(–1,–2)

   $(y - k)^2 = 4p(x - h)$

   $(y - -2)^2 = 4(-2)(x - -1)$

   $(y + 2)^2 = -8(x + 1)$

5. $x^2 = 8y$, $4p = 8$, $p = 2$

$F(0,2)$

directrix: $y = -2$

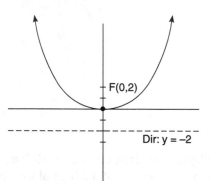

6. $y^2 - 4y - 4x = 0$

$(y^2 - 4y + 4) = 4x + 4$

$(y - 2)^2 = 4(x + 1)$ $4p = 4$, $p = 1$

$V(-1,2)$, $F(2,2)$, directrix: $x = 0$

7. $x^2 + 8x - y + 6 = 0$

$(x^2 + 8x + 16) = y - 6 + 16$

$(x + 4)^2 = y + 10$

$(x + 4)^2 = 1(y + 10)$ $4p = 1$, $p = \dfrac{1}{4}$

$V(-4,-10)$, $F\left(-4,-\dfrac{39}{4}\right)$, directrix: $y = -\dfrac{41}{4}$

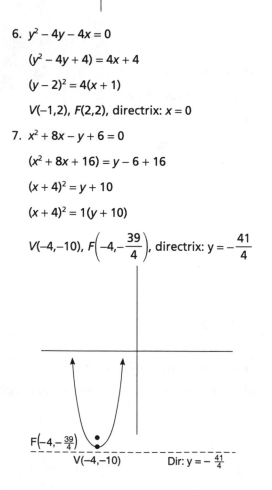

8. The center is (0,0). The major axis is on the *y*-axis because the y^2 term has the larger denominator.

   $a = 7, b = 5, c = \sqrt{24} = \pm 2\sqrt{6}$

   The vertices are (0,7) and (0,–7). The foci are $(0, 2\sqrt{6})$ and $(0, -2\sqrt{6})$.

9. $4x^2 + y^2 - 8x + 4y - 8 = 0$ Begin by completing the square.

   $4(x^2 - 2x + 1) + (y^2 + 4y + 4) = 8 + 4 + 4$

   $\dfrac{(x-1)^2}{4} + \dfrac{(y+2)^2}{16} = 1$ The vertices are (1,–6) and (1,2).

   The foci are $(1, -2 - 2\sqrt{3})$ and $(1, -2 + 2\sqrt{3})$.

10. $\dfrac{(y-2)^2}{4} - \dfrac{(x-3)^2}{9} = 1$

11. center: (0, 2)   12. $4\sqrt{2}$

    vertices: (–6,2), (6,2)

    foci: $(\pm 2\sqrt{10}, 2)$

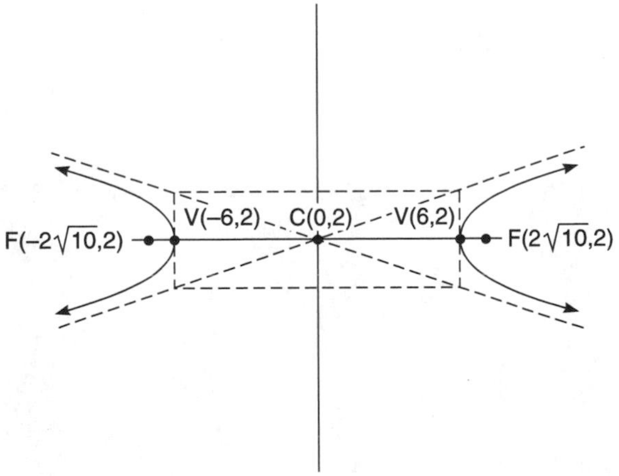

13. a. 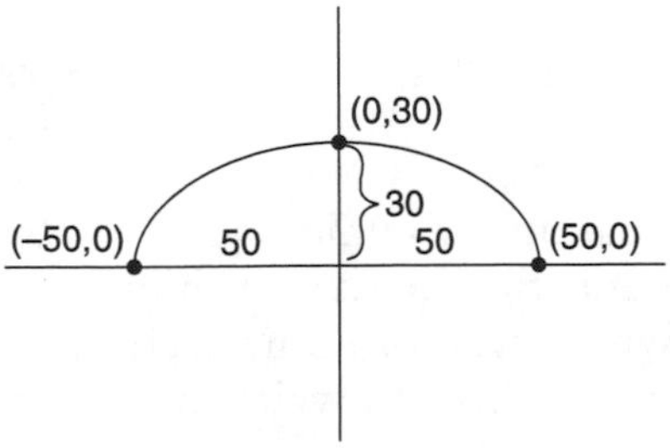   b. $\dfrac{x^2}{2500} + \dfrac{y^2}{900} = 1$ c. 13.08 feet

14. 2.083 feet from the vertex   15. 0 feet, 12.99 feet, 15 feet, 12.99 feet, 0 feet

A conic section is formed by intersecting a plane with a double-napped cone. The following illustration shows the various types of conic sections that can be formed. The point and the line are called degenerate conics. Our precalculus self-teaching guide (also published by Wiley) covers points and lines, so here we'll teach you about circles, ellipses, parabolas, and hyperbolas.

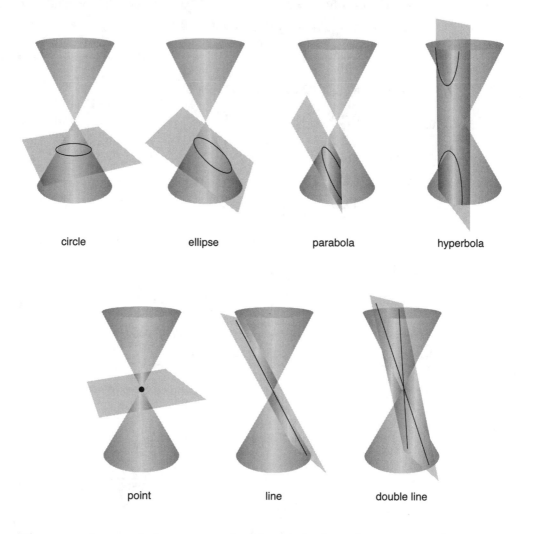

circle   ellipse   parabola   hyperbola

point   line   double line

The general second-degree equation for conics is $Ax^2 + Bxy + Cy^2 + Dx + Ey + F = 0$. This is a second-degree equation because its highest exponent is 2. We will show you how to recognize which type of conic section an equation represents, how to graph the conic section, and how to write the equation of a conic section given the graph, then we'll show you some applications of conic sections. Let's get started with circles.

Circles

Definition of a Circle

As you might recall, *a circle is the set of all points in a plane that are equidistant from a fixed point called the center.* The fixed distance from the circle's center to any point on the circumference of the circle is called the radius.

The following illustration shows a circle of radius r and center point (h,k), along with one point on the circumference of the circle itself (x_1,y_1).

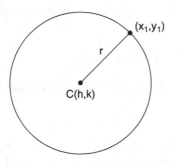

It shows two points that are on the radius of the circle. The center point (h,k) and a point on the circle's circumference (x,y). The distance from one of these points to the other is the radius r of the circle. We can easily use the distance formula to write an equation to represent this relationship. Just in case you forgot the distance formula, we'll take this opportunity to remind you that the distance between two points (x_1,y_1) and (x_2,y_2) connected by a straight line is $d = \sqrt{(x_2 - x_1)^2 + (y_2 - y_1)^2}$. If we apply this formula to the points on the circle (h,k) and (x,y), and substitute r for d, we have $\sqrt{(x - h)^2 + (y - k)^2} = r$. Next we'll square both sides of the equation to drop the radical. We get $(x - h)^2 + (y - k)^2 = r^2$.

Standard Form of the Equation of a Circle

Given a point (x,y) on the circumference of a circle with radius r and center point (h,k),

Equation of a circle (standard form):
$$(x - h)^2 + (y - k)^2 = r^2$$

Equation of a circle (general form):
$$Ax^2 + Bxy + Cy^2 + Dx + Ey + F = 0$$

The general form is obtained by expanding the standard form of a circle. It can be used to describe, mathematically, any circle. By using this equation, eventually we will be able to place the circle on a graph.

Example 1:

Write the equation of the circle with center point (2,–1) and radius 3. Write the equation in standard form and general form.

Solution:

We'll use the equation $(x - h)^2 + (y - k)^2 = r^2$ and substitute the given values for h, k, and r.

$(x - h)^2 + (y - k)^2 = r^2$ Substitute 2 for h, −1 for k, and 3 for r.

$(x - 2)^2 + (y - -1)^2 = 3^2$ Make the double negative a positive and square the 3.

$(x - 2)^2 + (y + 1)^2 = 9$

The equation of the circle in standard form is $(x - 2)^2 + (y + 1)^2 = 9$.

     To find the equation of the circle in general form, expand the binomials and combine like terms.

$(x - 2)^2 + (y + 1)^2 = 9$ Write the binominals twice.

$(x - 2)(x - 2) + (y + 1)(y + 1) = 9$ Distribute.

$x^2 - 2x - 2x + 4 + y^2 + y + y + 1 = 9$ Combine like terms in descending order.

$x^2 + y^2 - 4x + 2y - 4 = 0$

The equation of the circle in general form is $x^2 + y^2 - 4x + 2y - 4 = 0$.

Example 2:

Write the equation of the circle with center (1,0) that has a point on its circumference at (5,3). Write the equation in standard form and in general form.

Solution:

We're given the center of the circle but not its radius. Before we can write the equation of the circle, we have to find its radius. In order to find the radius of the circle, we have to calculate the distance from the center point (1,0) and a point on its circumference. We know (5,3) is on the circumference of the circle, so we'll apply the distance formula to the points (1,0) and (5,3).

     Again, the distance between two points (x_1, y_1) and (x_2, y_2) connected by a straight line is $d = \sqrt{(x_2 - x_1)^2 + (y_2 - y_1)^2}$.

We'll substitute 5 for x_2, 1 for x_1, 3 for y_2, and 0 for y_1. Now we're ready to find the radius of the circle.

$r = \sqrt{(5-1)^2 + (3-0)^2}$ Replace $(5-1)$ by 4 and $(3-0)$ by 3.

$r = \sqrt{(4)^2 + (3)^2}$ Square 4 and square 3.

$r = \sqrt{16 + 9}$ Add 16 and 9.

$r = \sqrt{25} = 5$ The radius of the circle is 5.

Now that we know the length of the radius of the circle, we can substitute 5 for r, 1 for h, and 0 for k into the standard formula of a circle.

$(x-1)^2 + (y-0)^2 = 5^2$ The equation of this circle in standard form is $(x-1)^2 + y^2 = 25$.

To write the general form of this circle, we expand $(x-1)^2$ and combine the like terms.

$(x-1)^2 = (x-1)(x-1) = x^2 - x - x + 1 = x^2 - 2x + 1$

$x^2 - 2x + 1 + y^2 = 25$ Subtract 25 from both sides of the equation.

$x^2 + y^2 - 2x + 1 - 25 = 0$

$x^2 + y^2 - 2x - 24 = 0$

The equation of this circle in general form is $x^2 + y^2 - 2x - 24 = 0$.

Example 3:
Write the equation of the circle with points (2, 3) and (−1, −2) on its diameter and its circumference. Write the equation in standard form and general form.

Solution:
In order to write the equation of the circle, we have to find its radius and its center point. Since we know the given points are on the diameter and also on the circumference of the circle, we know the midpoint of the line that connects these points would have to be the center of the circle and that half the length of the line that connects these points is the radius of the circle. We'll start by reminding you of the formula for the midpoint of a line segment with end points (x_2, y_2) and (x_1, y_1):

$$Midpoint = \left(\frac{x_2 + x_1}{2}, \frac{y_2 + y_1}{2} \right)$$

The center point of this circle is $C = \left(\dfrac{-1 + 2}{2}, \dfrac{-2 + 3}{2} \right) = \left(\dfrac{1}{2}, \dfrac{1}{2} \right)$.

Now that we know the center point of the circle, all we have to do is use the distance formula to find the diameter of the circle and take half of that to find the radius of the circle.

$$d = \sqrt{(x_2 - x_1)^2 + (y_2 - y_1)^2} = \sqrt{(-1 - 2)^2 + (-2 - 3)^2} = \sqrt{(-3)^2 + (-5)^2}$$
$$= \sqrt{9 + 25} = \sqrt{34}$$

The diameter is $\sqrt{34}$ so the radius is $\dfrac{1}{2}\sqrt{34}$.

Now we'll substitute $\dfrac{1}{2}$ for h, $\dfrac{1}{2}$ for k, and $\dfrac{1}{2}\sqrt{34}$ for r into $(x - h)^2 + (y - k)^2 = r^2$ to write the standard form of this circle.

$$\left(x - \frac{1}{2}\right)^2 + \left(y - \frac{1}{2}\right)^2 = \left(\frac{1}{2}\sqrt{34}\right)^2 \qquad \text{Let's simplify.} \left(\frac{1}{2}\sqrt{34}\right)^2 = \frac{1}{4}(34) = \frac{17}{2}$$

The standard form of the equation of this circle is $\left(x - \dfrac{1}{2}\right)^2 + \left(y - \dfrac{1}{2}\right)^2 = \dfrac{17}{2}$.

To write this in general form, we'll expand the binominals and combine like terms.

$$\left(x - \frac{1}{2}\right)^2 + \left(y - \frac{1}{2}\right)^2 = \frac{17}{2} \qquad \text{Write the binominals twice.}$$

$$\left(x - \frac{1}{2}\right)\left(x - \frac{1}{2}\right) + \left(y - \frac{1}{2}\right)\left(y - \frac{1}{2}\right) = \frac{17}{2} \qquad \text{Distribute.}$$

$$x^2 - \frac{1}{2}x - \frac{1}{2}x + \frac{1}{4} + y^2 - \frac{1}{2}y - \frac{1}{2}y + \frac{1}{4} = \frac{17}{2} \qquad \text{Combine like terms.}$$

$$\frac{1}{4} + \frac{1}{4} - \frac{17}{2} = \frac{1}{2} - \frac{17}{2} = -\frac{16}{2} = -8$$

$$-\frac{1}{2}x - \frac{1}{2}x = -\frac{2}{2}x = -1x = -x$$

$$-\frac{1}{2}y - \frac{1}{2}y = -\frac{2}{2}y = -1y = -y$$

$$x^2 - x + y^2 + y - 8 = 0$$
$$x^2 + y^2 - x - y - 8 = 0$$

The general form of this circle is $x^2 + y^2 - x - y - 8 = 0$.

Example 4:
Write the equation of the circle with center point (0,0) and radius *r*.

Solution:
Substitute 0 for *h*, 0 for *k*, and *r* for *r*.

$(x - h)^2 + (y - k)^2 = r^2$

$(x - 0)^2 + (y - 0)^2 = r^2$ $\qquad x - 0 = x, \ y - 0 = y$

$x^2 + y^2 = r^2$

Equation of any circle with radius r centered at the origin.
$x^2 + y^2 = r^2$

SELF-TEST 1

1. Write the equation in standard form of the circle with radius 10 and center point (2,−5).

2. Write the equation in general form of the circle with radius 3 and center point (−1,4).

3. Write the equation of the circle in general form and standard form whose center point is (−2,4) and which has a point (3,5) on its circumference.

4. Write the equation of the following circle in standard form.

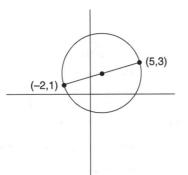

5. Write the equation in standard form and general form of the circle that has points (0,−3) and (9,2) as end points on its diameter.

6. Write the equation of the circle centered at the origin with radius 6.

ANSWERS

1. $(x - h)^2 + (y - k)^2 = r^2$ Substitute 10 for r, 2 for h, and -5 for k.

   $(x - 2)^2 + (y - -5)^2 = 10^2$ Change the double negative to a positive and square the 10.

   $(x - 2)^2 + (y + 5)^2 = 100$

2. $(x - h)^2 + (y - k)^2 = r^2$ Substitute 3 for r, -1 for h, and 4 for k.

   $(x - -1)^2 + (y - 4)^2 = 3^2$ Change the double negative to a positive and square the 3.

   $(x + 1)^2 + (y - 4)^2 = 9$ This is the standard form of the equation of this circle.

      To find the general form, we have to expand the binominals and combine the like terms.

   $(x + 1)^2 + (y - 4)^2 = 9$ Write the binominals twice.

   $(x + 1)(x + 1) + (y - 4)(y - 4) = 9$ Distribute.

   $x^2 + x + x + 1 + y^2 - 4y - 4y + 16 = 9$ Combine like terms.

   $x^2 + y^2 + 2x - 8y + 8 = 0$ This is the general form of the circle.

3. We're given the center point $(-2,4)$ but not the radius. We'll find the length of the radius of the circle by using the distance formula to find the distance from the center point $(-2,4)$, to given point $(3,5)$, which is on the circumference of the circle.

   $d = \sqrt{(x_2 - x_1)^2 + (y_2 - y_1)^2}$

   $r = \sqrt{(-2 - 3)^2 + (4 - 5)^2} = \sqrt{(-5)^2 + (-1)^2} = \sqrt{25 + 1} = \sqrt{26}$

   Now that we know the measure of the radius, all we have to do is substitute -2 for h, 4 for k, and $\sqrt{26}$ for r.

   $(x - h)^2 + (y - k)^2 = r^2$

   $(x - -2)^2 + (y - 4)^2 = (\sqrt{26})^2$ Change the double negative to a positive

   $(x + 2)^2 + (y - 4)^2 = 26$ and replace $(\sqrt{26})^2$ with 26.

   The standard form of this circle is $(x + 2)^2 + (y - 4)^2 = 26$. To find the general form of this circle, we'll expand the binomials and combine the like terms.

   $(x + 2)^2 + (y - 4)^2 = 26$ Write the binomials twice.

   $(x + 2)(x + 2) + (y - 4)(y - 4) = 26$ Distribute.

   $x^2 + 2x + 2x + 4 + y^2 - 4y - 4y + 16 = 26$ Combine the like terms.

   $x^2 + y^2 + 4x - 8y - 6 = 0$

   The general form of this circle is $x^2 + y^2 + 4x - 8y - 6 = 0$.

4. We'll begin by using the midpoint formula to find the center point of the circle.

$$Md = \left(\frac{x_2 + x_1}{2}, \frac{y_2 + y_1}{2}\right)$$

$$Md = \left(\frac{5 - 2}{2}, \frac{3 + 1}{2}\right) = \left(\frac{3}{2}, 2\right)$$

Now that we know the center point of the circle, we'll use half the distance formula to find the length of the radius.

$d = \sqrt{(x_2 - x_1)^2 + (y_2 - y_1)^2}$　　　　　Substitute in the values for the points.

$r = \frac{1}{2}\sqrt{(5 - -2)^2 + (3 - 1)^2}$　　　　　Change the double negative to a positive.

$r = \frac{1}{2}\sqrt{(5 + 2)^2 + (3 - 1)^2}$　　　　　Combine the numbers.

$r = \frac{1}{2}\sqrt{(7)^2 + (2)^2}$　　　　　Square the numbers.

$r = \frac{1}{2}\sqrt{49 + 4}$　　　　　Add the numbers.

$r = \frac{1}{2}\sqrt{53}$

Now that we know the coordinates of the center point and the length of the radius, we'll substitute these values into $(x - h)^2 + (y - k)^2 = r^2$ to write the equation of the circle in standard form.

$(x - h)^2 + (y - k)^2 = r^2$　　　　　Substitute $\frac{3}{2}$ for h, 2 for k, and $\frac{1}{2}\sqrt{53}$ for r.

$$\left(x - \frac{3}{2}\right)^2 + (y - 2)^2 = \left(\frac{1}{2}\sqrt{53}\right)^2 \qquad \left(\frac{1}{2}\sqrt{53}\right)\left(\frac{1}{2}\sqrt{53}\right) = \frac{1}{4}(53) = \frac{53}{4}$$

$$\left(x - \frac{3}{2}\right)^2 + (y - 2)^2 = \frac{53}{4}$$

The standard equation of this circle is $\left(x - \frac{3}{2}\right)^2 + (y - 2)^2 = \frac{53}{4}$.

5. First we'll find the midpoint of the diameter to find the center point of the circle.

$$Md = \left(\frac{x_2 + x_1}{2}, \frac{y_2 + y_1}{2}\right) \quad \text{Substitute the coordinates of the given points.}$$

$$Md = \left(\frac{0 + 9}{2}, \frac{-3 + 2}{2}\right) = \left(\frac{9}{2}, -\frac{1}{2}\right) \qquad \text{Now we know } h = \frac{9}{2} \text{ and } k = -\frac{1}{2}.$$

The center of this circle is $\left(\frac{9}{2}, -\frac{1}{2}\right)$.

All we need now is the radius of the circle and we're ready for anything. As we know, the radius is one-half the diameter.

$$d = \sqrt{(x_2 - x_1)^2 + (y_2 - y_1)^2}$$

$$r = \frac{1}{2}\sqrt{(0-9)^2 + (-3-2)^2}$$ Combine the numbers.

$$r = \frac{1}{2}\sqrt{(-9)^2 + (-5)^2}$$ Square the numbers.

$$r = \frac{1}{2}\sqrt{81 + 25}$$ Add the numbers.

$$r = \frac{1}{2}\sqrt{106}$$

Now that we know the center point and the radius, we'll substitute these values into the standard form of the equation of a circle.

$(x-h)^2 + (y-k)^2 = r^2$ Substitute $\frac{9}{2}$ for h, $-\frac{1}{2}$ for k, and $\frac{1}{2}\sqrt{106}$ for r.

$\left(x - \frac{9}{2}\right)^2 + \left(y - -\frac{1}{2}\right)^2 = \left(\frac{1}{2}\sqrt{106}\right)^2$ Change the double negative to a positive.

$\left(x - \frac{9}{2}\right)^2 + \left(y + \frac{1}{2}\right)^2 = \frac{53}{2}$

The standard form of the equation is $\left(x - \frac{9}{2}\right)^2 + \left(y + \frac{1}{2}\right)^2 = \frac{53}{2}$.

Now we'll find the general form of the equation by expanding the standard form.

$\left(x - \frac{9}{2}\right)^2 + \left(y + \frac{1}{2}\right)^2 = \frac{53}{2}$ Write the binomials twice.

$\left(x - \frac{9}{2}\right)\left(x - \frac{9}{2}\right) + \left(y + \frac{1}{2}\right)\left(y + \frac{1}{2}\right) = \frac{53}{2}$ Distribute.

$x^2 - \frac{9}{2}x - \frac{9}{2}x + \frac{81}{4} + y^2 + \frac{1}{2}y + \frac{1}{2}y + \frac{1}{4} = \frac{53}{2}$ Combine the like terms.

$x^2 + y^2 - 9x + y + \frac{82}{4} = \frac{53}{2}$ $\frac{82}{4} - \frac{53}{2} = \frac{84}{4} - \frac{106}{4} = -\frac{24}{4} = -6$

$x^2 + y^2 - 9x + y - 6 = 0$

The general form of the equation of this circle is $x^2 + y^2 - 9x + y - 6 = 0$.

6. Because the center point is the origin, we can use the formula from example 4.

$x^2 + y^2 = r^2$

The equation of this circle is $x^2 + y^2 = 36$.

In the previous examples, we gave you information about the circle and asked you to give us the equation of the circle. Now we're going to give you the equation of a circle and ask you to find the center point and the radius of the circle. Don't worry. All we're going to do now is reverse the process that we just did.

Example 5:

The equation of a circle is $(x - 7)^2 + (y + 10)^2 = 25$. Find the center point and the radius of the circle.

Solution:

Let's see how this equation $(x - 7)^2 + (y + 10)^2 = 25$ compares with the standard form of the equation of a circle $(x - h)^2 + (y - k)^2 = r^2$.

Notice the standard form has subtraction inside the parentheses, but the given equation has addition inside the second set of parentheses. This indicates that it must have originally been $(y - -10)^2$, so the center point is $(7, -10)$. According to the standard form of the equation of a circle, $25 = r^2$, so the radius is $r = \sqrt{25} = 5$.

In example 5 it was easy to find the center and radius of the circle because it was written in standard form. Now we're going to find the center and radius of some circles that are written in general form, $Ax^2 + Bxy + Cy^2 + Dx + Ey + F = 0$. This is a little bit more complicated. We have to use a method called completing the square to convert a circle from general form to standard form. As you saw in the last example, it's easy to find the center point and the radius of a circle once it's in standard form. In example 6 we'll walk you through the steps to complete the square.

Example 6:

Find the center point and the radius of the following circle.

$16x^2 + 16y^2 - 16x + 24y - 3 = 0$

Solution:

As you can see this circle is in general form, so we'll use the method of completing the square to convert it to standard form. To complete the square, the coefficients of x^2 and y^2 have to be 1.

$16x^2 + 16y^2 - 16x + 24y - 3 = 0$	Divide every term by 16.
$x^2 + y^2 - x + \dfrac{3}{2}y - \dfrac{3}{16} = 0$	Now the leading coefficients are 1.
	Group the x's and y's together and move the constant to the other side of the equation.
$(x^2 - x) + \left(y^2 + \dfrac{3}{2}y\right) = \dfrac{3}{16}$	Now take half the coefficient of the first power terms of x and y and square them. Then add these numbers to both sides of the equation.

Let's start with $-x$, which is assumed to be $-1x$. Half of -1 is $-\dfrac{1}{2}$; squared it is $\left(-\dfrac{1}{2}\right)^2 = \dfrac{1}{4}$. We'll add $\dfrac{1}{4}$ to both sides of the equation. Now let's look at the coefficient of $\dfrac{3}{2}y$, which is $\dfrac{3}{2}$. We'll take half of $\dfrac{3}{2}$, which is $\left(\dfrac{1}{2}\right)\left(\dfrac{3}{2}\right) = \dfrac{3}{4}$; squared it is $\left(\dfrac{3}{4}\right)^2 = \dfrac{9}{16}$. Next we'll add $\dfrac{9}{16}$ to both sides of the equation. As you'll soon see, this process will create perfect binominal squares for the x and y terms.

$$\left(x^2 - x + \frac{1}{4}\right) + \left(y^2 + \frac{3}{2}y + \frac{9}{16}\right) = \frac{3}{16} + \frac{1}{4} + \frac{9}{16}$$

Factor and add $\dfrac{3}{16} + \dfrac{4}{16} + \dfrac{9}{16} = \dfrac{16}{16} = 1$.

$$\left(x - \frac{1}{2}\right)^2 + \left(y + \frac{3}{4}\right)^2 = 1$$

Let's compare this to the standard equation of a circle.

$$(x - h)^2 + (y - k)^2 = r^2$$

Now it's easy to see the center point is $\left(\dfrac{1}{2}, -\dfrac{3}{4}\right)$ and the radius is 1.

Let's go back a couple of steps to where we factored. You may have had some trouble with this step, so let's go over it slowly. To factor the trinomial $\left(x^2 - x + \dfrac{1}{4}\right)$, we know we need x times x to get x^2, and then we need two numbers that will multiply to $\dfrac{1}{4}$ and add to equal -1. Those numbers are $-\dfrac{1}{2}$ and $-\dfrac{1}{2}$. A faster and easier way to figure this out when completing the square is to use half the coefficient of the first-power term, which is how we got the $-\dfrac{1}{2}$ in the first place. Another way is simply to take the square root of the constant, $\sqrt{\dfrac{1}{4}} = \dfrac{1}{2}$, and keep the sign of the middle term. The same is true of $\left(y^2 + \dfrac{3}{2}y + \dfrac{9}{16}\right)$. Half the middle term is $\dfrac{3}{4}$. Or we could simply say $\sqrt{\dfrac{9}{16}} = \dfrac{3}{4}$, and keep the sign of the middle term. Once we've factored $\left(x^2 - x + \dfrac{1}{4}\right)$ to $\left(x - \dfrac{1}{2}\right)^2$ and $\left(y^2 + \dfrac{3}{2}y + \dfrac{9}{16}\right)$ to $\left(y + \dfrac{3}{4}\right)^2$, we've created perfect squares that make it easy for us to find the center point of the circle. The following is an outline of how to complete the square.

Steps for Completing the Square

1. Both coefficients of the square terms must be 1. Multiply or divide to make them 1.

2. Rewrite the equation grouping the terms that have an x together in descending order. Do the same for the terms that have a y. Move the constant to the other side of the equation.

3. Square half the coefficient of the first power terms, and add these numbers to both sides of the equation.

4. Factor.

Let's see how good you are at completing the square. Try the following problem, then check your answer with ours.

Example 7:

Find the center and radius of the following circle.
$4x^2 + 4y^2 - 16y + 15 = 0$

Solution:

The leading coefficients are not 1 so we'll begin by dividing all terms of the equation by 4.

$4x^2 + 4y^2 - 16y + 15 = 0$ — Divide by 4.

$x^2 + y^2 - 4y + \dfrac{15}{4} = 0$ — Group like terms together and move the constant to the other side of the equation.

$x^2 + y^2 - 4y = -\dfrac{15}{4}$ — Add the square of half the coefficient of the first power terms to both sides of the equation. There isn't an x to the first power term, so we won't worry about the x. The coefficient of the y to the first power term is -4. Half of -4 is -2; squared it is 4. x^2 can be written as $(x-0)^2$.

$(x-0)^2 + (y^2 - 4y + 4) = -\dfrac{15}{4} + 4$ — Factor, then add: $-\dfrac{15}{4} + 4 = -\dfrac{15}{4} + \dfrac{16}{4} = \dfrac{1}{4}$

$(x-0)^2 + (y-2)^2 = \dfrac{1}{4}$ — The center point is $(0,2)$. $r^2 = \dfrac{1}{4}$. To find the radius, just take the square root of

$\dfrac{1}{4}$. $r = \sqrt{\dfrac{1}{4}} = \dfrac{1}{2}$.

Example 8:

Find the center and radius of the following circle.

$x^2 + y^2 = 8$

Solution:

This example is a little different from the previous ones. It doesn't have any terms to the first power. There's no point in completing the square for this one. We'll just rewrite the equation.

$(x - 0)^2 + (y - 0)^2 = 8$ The center is $(0,0)$. $r^2 = 8$. To find the value of r, just take the square root of 8. $r = \sqrt{8} = \sqrt{4(2)} = 2\sqrt{2}$.

If you need to review simplifying radicals or completing the square, try our precalculus self-teaching guide, which is also published by Wiley.

Example 9:

Write the following circle in standard form and graph. Label four points on the circumference of the circle.

$x^2 + y^2 + 4x - 6y - 23 = 0$

Solution:

First we'll use the method of completing the square to write the given circle in standard form so we can find the center point and radius of the circle.

$x^2 + y^2 + 4x - 6y - 23 = 0$ The leading coefficients are both 1 so we don't have to multiply or divide the equation by a constant.

$(x^2 + 4x) + (y^2 - 6y) = 23$ Group the same variables together.

$(x^2 + 4x + 4) + (y^2 - 6y + 9) = 23 + 4 + 9$ Complete the square, then factor.

$(x + 2)^2 + (y - 3)^2 = 36$ The center point is $(-2,3)$ and the radius is 6.

With this information, it becomes easy to find other points on the circle. The easiest way to find four points on the circumference of the circle is to begin at the center point $(-2,3)$ and use the knowledge that the radius is 6 units. If we move horizontally (left or right) from the center point, the x coordinate of the center point changes. If we move vertically (up or down) from the center point, the y coordinate of the center point changes. Moving from the center point $(-2,3)$ to the right 6 units, we get the point $(-2 + 6,3)$, which is $(4,3)$. Moving from the center point $(-2,3)$ to the left 6 units, we get the point $(-2 - 6,3)$, which is

(−8,3). Moving from the center point (−2,3) up 6 units, we get the point (−2,3 + 6), which is (−2,9). Finally, moving from the center point (−2,3) down 6 units, we get the point (−2,3 − 6) which is (−2,−3).

Four of the points on the circumference of the circle are: (4,3), (−8,3), (−2,9), and (−2,−3).

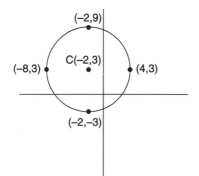

SELF-TEST 2

1. Find the center and radius of the following circle: $x^2 + y^2 - 4x + 8y + 11 = 0$

   Graph the circle and label four points on the circumference of the circle.

2. Find the center and radius of the following circle: $2x^2 + 2y^2 + 12x - 8y - 31 = 0$

3. Find the center and radius of the following circle: $x^2 + y^2 - 6y - 7 = 0$

4. Find the center and radius of the following circle: $36x^2 + 36y^2 - 48x + 36y - 119 = 0$

5. Find the center and radius of the following circle: $x^2 + y^2 = 48$

ANSWERS

1. $x^2 + y^2 - 4x + 8y + 11 = 0$

   The leading coefficients are 1 so we don't have to multiply or divide the equation by a constant.

   $(x^2 - 4x) + (y^2 + 8y) = -11$

   Group the same variables together and move the constant to the other side of the equation.

   $(x^2 - 4x + 4) + (y^2 + 8y + 16) = -11 + 4 + 16$

   Complete the square on the binominals.

   $(x - 2)^2 + (y + 4)^2 = 9$

   Now the equation is in standard form.

The center point is (2,–4) and the radius is 3.

To easily find four points on the circumference of the circle, we'll begin at the center point (2,–4) and move to the right three units, left three units, up three units, and down three units. Moving to the right or left will change the x coordinate of the center point; moving up or down will change the y coordinate of the center point. The points we'll label on the circumference are (5,–4), (–1,–4), (2,–1), and (2,–7).

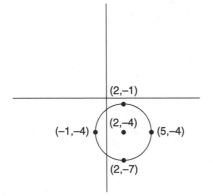

2. $2x^2 + 2y^2 + 12x - 8y - 31 = 0$

We'll begin by dividing by 2 to change the leading coefficients to 1.

$x^2 + y^2 + 6x - 4y - \dfrac{31}{2} = 0$

Group the same variables together and move the constant to the other side of the equation.

$(x^2 + 6x) + (y^2 - 4y) = \dfrac{31}{2}$

Complete the square on the binomials.

$(x^2 + 6x + 9) + (y^2 - 4y + 4) = \dfrac{31}{2} + 9 + 4$

Factor and add: $\dfrac{31}{2} + \dfrac{18}{2} + \dfrac{8}{2} = \dfrac{57}{2}$.

$(x + 3)^2 + (y - 2)^2 = \dfrac{57}{2}$

Now the equation is in standard form.

The center is (–3,2) and the radius

$r = \sqrt{\dfrac{57}{2}} = \dfrac{\sqrt{57}}{\sqrt{2}}\left(\dfrac{\sqrt{2}}{\sqrt{2}}\right) = \dfrac{\sqrt{114}}{2}$

3. $x^2 + y^2 - 6y - 7 = 0$

The leading coefficients are already 1.

$x^2 + (y^2 - 6y) = 7$

Group the same variables together and move the constant to the other side of the equation.

$x^2 + (y^2 - 6y + 9) = 7 + 9$

Complete the square for the y's.

$(x + 0)^2 + (y - 3)^2 = 16$

The equation is in standard form.

The center is (0,3) and the radius is 4.

4. $36x^2 + 36y^2 - 48x + 36y - 119 = 0$

We'll begin by dividing every term of the equation by 36 to change the leading coefficients to 1.

$$x^2 + y^2 - \frac{48}{36}x + y - \frac{119}{36} = 0$$

$$\left(x^2 - \frac{4}{3}x\right) + (y^2 + y) = \frac{119}{36}$$

Complete the square for both binomials.

$$\left(x^2 - \frac{4}{3}x + \frac{4}{9}\right) + \left(y^2 + y + \frac{1}{4}\right) = \frac{119}{36} + \frac{4}{9} + \frac{1}{4}$$

$$\left[\frac{1}{2}\left(-\frac{4}{3}\right)\right]^2 = \left[-\frac{2}{3}\right]^2 = \frac{4}{9}$$

$$\left(x - \frac{2}{3}\right)^2 + \left(y + \frac{1}{2}\right)^2 = 4$$

$$\left[\frac{1}{2}(1)\right]^2 = \left[\frac{1}{2}\right]^2 = \frac{1}{4}$$

$$\frac{119}{36} + \frac{4}{9} + \frac{1}{4} = \frac{119}{36} + \frac{16}{36} + \frac{9}{36} = \frac{144}{36} = 4$$

The center is $\left(\frac{2}{3}, -\frac{1}{2}\right)$ and the radius is 2.

5. $x^2 + y^2 = 48$

This equation doesn't have any first power terms, so we don't have to complete the square. It's of the form $x^2 + y^2 = r^2$.

The center point is (0, 0) and the radius $r = \sqrt{48} = \sqrt{16(3)} = 4\sqrt{3}$.

Parabolas

In algebra we study parabolas as graphs of quadratic equations. They either open up or down, right or left. In this section we will use a geometric definition of a parabola to write the equation of the parabola. We will also give you the equation of a parabola and ask you to find its focus, vertex, and directrix and sketch its graph.

A parabola is the set of all points (x,y) that are equidistant from a fixed line, the directrix, and a fixed point, the focus, not on the line. See the following figure.

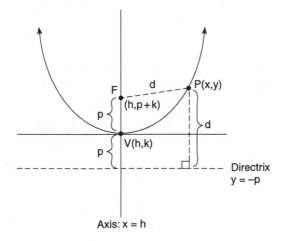

The midpoint between the focus and directrix is the vertex. The line that passes through the focus and the vertex is the axis of the parabola. The distance from the focus to the vertex is equal to the distance from the vertex to the directrix. We'll refer to this distance as p. The previous figure has a vertical axis.

The vertex of a parabola doesn't have to be at the origin $(0,0)$, just as the center of a circle in the last section didn't have to be at the origin. Just to make things a little easier for you, we'll begin this section by working on parabolas that do have a vertex at the origin. Once you have the hang of that, we'll show you how to work on parabolas that have a vertex that is not at the origin.

We'll start by showing you how to derive the formula of a parabola with a vertex at $(0,0)$ and a focus p units from the vertex and p units from the directrix.

Based on the definition of a parabola, we can assume the distance from a point, P, on the parabola, to a point F, the focus of the parabola, is equal to the distance from P to a specific point, D, on the directrix. Using our old friend the distance formula, let's write that in an equation form. Just in case you forgot the distance formula, we'll take this opportunity to remind you that the distance between two points (x_1, y_1) and (x_2, y_2) connected by a straight line is $d = \sqrt{(x_2 - x_1)^2 + (y_2 - y_1)^2}$. We'll apply this formula to the following figure.

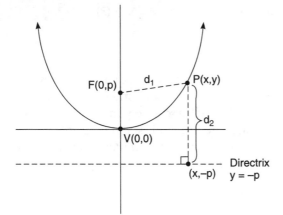

$d_1 = d_2$

Assume the distances are equal and use the distance formula on each side of the equation.

$$\sqrt{(x - 0)^2 + (y - p)^2} = \sqrt{(x - x)^2 + (y - -p)^2}$$

$$(x - 0)^2 + (y - p)^2 = (y + p)^2$$

We squared both sides of the equation to drop the radical.

$$x^2 + y^2 - 2py + p^2 = y^2 + 2py + p^2$$

Combine the like terms and isolate the x^2.

$$x^2 = 4py$$

This is the standard form of the equation of a parabola with vertex $(0,0)$ and a vertical axis.

Now we'll derive the formula of a parabola with vertex (0,0) and a horizontal axis. Refer to the following figure as we work our way through the derivation of the formula.

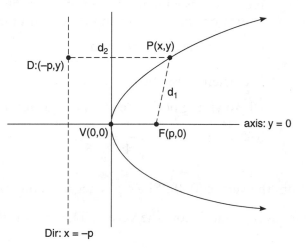

$d_1 = d_2$ — Assume the distances are equal and use the distance formula.

$\sqrt{(x-p)^2 + (y-0)^2} = \sqrt{(-p-x)^2 + (y-y)^2}$ — Square both sides to drop the radical.

$(x-p)^2 + y^2 = (-p-x)^2$ — Distribute.

$x^2 - 2px + p^2 + y^2 = p^2 + 2px + x^2$ — Combine like terms and isolate the y^2.

$y^2 = 4px$ — This is the standard form of the equation of a parabola with vertex (0,0) and a horizontal axis.

Notice an important fact: Unlike the equation of a circle, the equation of a parabola does not have both the x and the y squared. If the equation has an x^2, the parabola either opens up or down. If the equation has a y^2, the parabola either opens to the right or the left. In a little while we'll discuss how we know which way a given parabola opens: up or down, right or left.

Example 10:
Find the focus of the parabola whose equation is $y = -2x^2$.

Solution:
The x is squared, not the y, so we know this parabola has a vertical axis and therefore either opens up or down, not to the right or left: Its standard form is $x^2 = 4py$. We can manipulate the given equation to fit the standard form.

$y = -2x^2$

Divide both sides of the equation by -2, or multiply by $-\dfrac{1}{2}$.

$-\dfrac{1}{2}y = x^2$

Rewrite this equation with the x^2 on the left and compare it to $x^2 = 4py$. In the standard form, $4p$ is the coefficient of y. In the equation $x^2 = -\dfrac{1}{2}y$, $-\dfrac{1}{2}$ is the coefficient of y. We can assume $4p = -\dfrac{1}{2}$. Solve for p by dividing both sides of the equation by 4.

$x^2 = -\dfrac{1}{2}y$

$p = -\dfrac{1}{2} \div 4 = -\dfrac{1}{2} \times \dfrac{1}{4} = -\dfrac{1}{8}$

The focus is p units from the vertex $(0, 0)$. Since p is negative, the parabola opens down; therefore, the focus is p units below the vertex. The focus is $\left(0, -\dfrac{1}{8}\right)$.

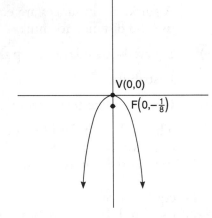

Example 11:

Write the standard form of the equation of the parabola with vertex (0,0) and focus (–2,0). Also write the equation of the directrix.

Solution:

First we have to figure out whether the parabola has a horizontal or a vertical axis. Because the coordinates of the vertex and focus have the same y value, we know the parabola opens to the left or the right. The focus is inside the parabola, so we know this one opens to the left. It will help to draw the parabola.

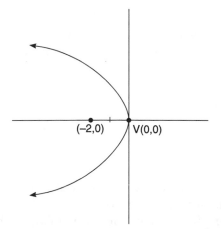

Because the focus is p units from the vertex, we know $p = -2$. The -2 is the x-coordinate of the focus, so we know the focus is two units to the left of the vertex. The line that passes through the focus and vertex is the axis of the parabola. In this case the axis is horizontal. The directrix is perpendicular to the axis and is p units from the vertex in the opposite direction of the focus. The directrix is $x = 2$. The parabola opens to the left because p is negative and the focus and vertex are on a horizontal line. We'll use the form $y^2 = 4px$ because this is the form used when there is a horizontal axis.

$y^2 = 4px$ Substitute -2 for p and simplify.

$y^2 = 4(-2)x$

$y^2 = -8x$ This is the equation of the parabola with vertex $(0,0)$ and focus $(-2,0)$.

Example 12:

Write the equation of the following parabola in standard form.

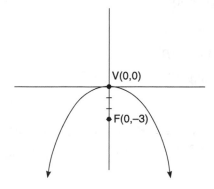

Solution:

The parabola opens down, so we know p is negative. We also know the x is squared, not the y. The form we have to use is $x^2 = 4py$. The focus is 3 units below the vertex, so $p = -3$.

$x^2 = 4py$	Substitute -3 for p.
$x^2 = 4(-3)y$	Multiply $4(-3)$.
$x^2 = -12y$	

So far all the parabolas we've worked on had a vertex at $(0,0)$. Now it's time to make things a little more challenging. We're going to work on parabolas in which the vertices are not at the origin. We have new formulas for this case.

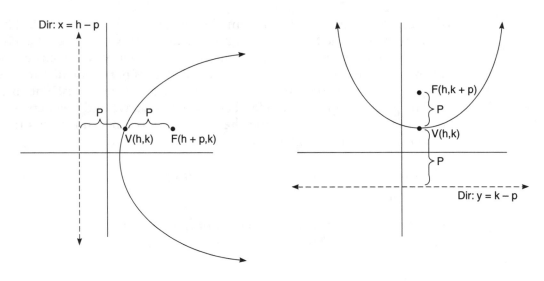

Left or Right Opening

$(y - k)^2 = 4p(x - h)$

Vertex: (h,k)

Focus: $(h + p, k)$

Directrix: $x = h - p$

Up or Down Opening

$(x - h)^2 = 4p(y - k)$

Vertex: (h,k)

Focus: $(h, k + p)$

Directrix: $y = k - p$

Example 13:

Graph the following parabola. Label the vertex, focus, and directrix of the parabola. Label two other points on the parabola.

$(x - 5)^2 = -2(y + 1)$

Solution:

Because the x is squared, not the y, we know the graph opens either up or down. If p is positive, the parabola opens up. If p is negative, the parabola opens down. Next we'll find the value of p.

$(x - h)^2 = 4p(y - k)$

$(x - 5)^2 = -2(y + 1)$

$4p = -2$ Divide both sides of the equation by 4.

$p = -\dfrac{2}{4} = -\dfrac{1}{2}$ p is negative, so the parabola opens down.

The vertex is $(5, -1)$. To find the focus, we have to add $-\dfrac{1}{2}$ to the y coordinate of the vertex. The focus is $\left(5, -\dfrac{3}{2}\right)$. The directrix is p units from the vertex in the opposite direction from the focus. The directrix is $y = -\dfrac{1}{2}$. This is the first example where we are going to take the information we found and graph the equation. To find two other points on the parabola, we're going to look at the given equation, $(x - 5)^2 = -2(y + 1)$. We notice that the left side of the equation is a perfect square; the right side is not. What value(s) can we substitute in for y that would make the right side of the equation a perfect square? If the right side of the equation was 4, for example, the equation would have perfect squares on both sides. In order for the right side of the equation to be a 4, $-2(y + 1)$ would have to equal 4. What value of y would make $-2(y + 1)$ equal 4?

$-2(y + 1) = 4$ Distribute -2.

$-2y - 2 = 4$ Add 2 to both sides of the equation.

$-2y = 6$ Divide both sides of the equation by -2.

$y = -3$

Substitute -3 for y in $(x - 5)^2 = -2(y + 1)$.

$(x - 5)^2 = -2(-3 + 1)$ Add the -3 and the 1.

$(x - 5)^2 = -2(-2)$ Multiply $2(-2) = 4$

$(x - 5)^2 = 4$ Since both sides of the equation are perfect squares, take the square root of both sides.

$x - 5 = \pm 2$ Split this into two equations and solve for x.

$x - 5 = 2, x - 5 = -2$

$x = 7, x = 3$

Now we know two points on the parabola are $(7,-3)$, and $(3,-3)$.

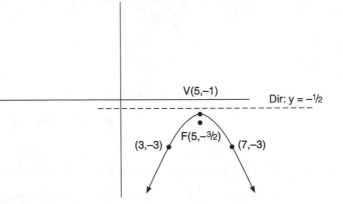

Now it's your turn to try a few problems.

SELF-TEST 3

1. Find the vertex, directrix, and focus of the following parabola.

   $y = 2x^2$

2. Find the vertex, directrix, and focus of the following parabola. Does is open up, down, to the right, or to the left?

   $y^2 = -\dfrac{1}{4}x$

3. Write the equation of the parabola with vertex $(0,0)$ and focus $(0,-5)$.

4. Write the equation of the parabola with focus at $(-2,0)$ and directrix the line $x = 2$.

5. Graph the following parabola. Label the vertex, directrix, focus, and two other points on the parabola.

   $(x - 2)^2 = 8(y + 3)$

ANSWERS

1. $y = 2x^2$

   The x is squared, not the y, so we know it opens up or down. We have to rewrite the given equation in standard form, $x^2 = 4py$. To do this we'll divide both sides of the equation by 2 or multiply by $\dfrac{1}{2}$.

$\frac{1}{2}y = x^2$ The coefficient of y is $4p$.

$\frac{1}{2} = 4p$ Divide both sides of the equation by 4, which is the same as multiplying by $\frac{1}{4}$.

$\frac{1}{8} = p$ Since p is positive, we know the parabola opens up, not down.

The vertex is $(0,0)$. To find the focus, we add $\frac{1}{8}$ to the y coordinate of the vertex. The focus is $\left(0,\frac{1}{8}\right)$. The directrix is the horizontal line that is $\frac{1}{8}$ units from the vertex in the opposite direction from the focus. The directrix is $y = -\frac{1}{8}$.

2. $y^2 = -\frac{1}{4}x$ In this equation the y is squared, not the x, so we know the parabola will open to the right if p is positive and to the left if p is negative. The standard form of this type of parabola is $y^2 = 4px$. We'll begin by finding the value of p.

$4p = -\frac{1}{4}$ Divide both sides of the equation by 4.

$p = -\frac{1}{16}$ Since p is negative, we know the parabola will open to the left.

The vertex is $(0,0)$. The focus is p units from the vertex, or $\left(-\frac{1}{16},0\right)$. The directrix is a vertical line p units from the vertex in the opposite direction of the focus. The directrix is $x = \frac{1}{16}$.

3. The vertex is at the origin and the focus is 5 units below the vertex. This tells us that the parabola opens down; therefore, we know p is -5, and we use the formula $x^2 = 4py$. If we substitute -5 for p, we get the equation of the parabola.

$x^2 = 4py$ Substitute -5 for p.

$x^2 = 4(-5)y$ Multiply.

$x^2 = -20y$

4. The directrix is a vertical line, so the parabola has a y that's squared. Because the focus is to the left of the directrix, the parabola opens to the left and we can assume p is negative. The distance from the focus to the directrix is 4 units, so p is -2. We'll use all this information and the formula $(y - k)^2 = 4p(x - h)$ to write the equation of the parabola.

$(y - k)^2 = 4p(x - h)$ Substitute 0 for k, -2 for h, and -2 for p.

$(y - 0)^2 = 4(-2)(x - 0)$ Simplify the equation.

$y^2 = -8x$ This is the equation of the parabola with the given characteristics.

5. In the given equation, only the *x* is squared, not the *y*, so we know the parabola opens either up or down. Let's find the value for *p*.

$(x - h)^2 = 4p(y - k)$ $\qquad\qquad$ $4p = 8, p = 2$

$(x - 2)^2 = 8(y + 3)$ $\qquad\qquad$ $h = 2, k = -3$

V: (2,–3), F: (2,–1), *D:* *y* = –5

To find two other points on the parabola easily, we'll create a perfect square on the right side of the equation by letting $y = -1$.

$(x - 2)^2 = 8(-1 + 3)$ $\qquad\qquad$ Replace the $-1 + 3$ by 2.

$(x - 2)^2 = 8(2)$ $\qquad\qquad$ Replace 8(2) by 16.

$(x - 2)^2 = 16$ $\qquad\qquad$ Take the square root of both sides of the equation.

$x - 2 = \pm 4$ $\qquad\qquad$ Separate this into two equations and solve them for *x*.

$x - 2 = 4, x - 2 = -4$

$x = 6, x = -2$ $\qquad\qquad$ Two points on the parabola are (6,–1) and (–2,–1).

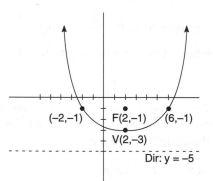

In the last section we had to complete the square on the equation of a circle given in general form to find the center and the radius of the circle. In this section we'll do the same to find the vertex and the value of *p*, given the equation of a parabola in standard form.

Example 14:

Find the vertex and value of *p* for the following parabola.

$y^2 + 2y - x = 0$

Solution:

The *y* is squared, not the *x*, so we'll complete the square on the *y* terms.

$y^2 + 2y - x = 0$ We'll begin by completing the square on the y terms and moving the x to the other side of the equal sign.

$y^2 + 2y = x$ Half the coefficient of the first power term is 1, squared is 1.

Add 1 to both sides of the equation.

$(y^2 + 2y + 1) = x + 1$ Factor.

$(y + 1)^2 = x + 1$

If we compare this equation to the standard form $(y - k)^2 = 4p(x - h)$, it's easy to see the vertex is at $(-1, -1)$. To find the value of p, we'll assume the $x + 1$ on the right side of the equation is $1(x - -1)$. Therefore $4p = 1$, and $p = \dfrac{1}{4}$.

Example 15:

Graph the following parabola. Label the vertex, focus, directrix, and two other points on the parabola.

$2x^2 - 8x - 4y + 3 = 0$

Solution:

The x is squared, not the y, so we'll complete the square on the x terms.

$2x^2 - 8x - 4y + 3 = 0$ Keep the x terms on one side and move all other terms to the other side of the equation.

$x^2 - 4x = 2y - \dfrac{3}{2}$ Divide all terms by 2 to make the leading coefficient 1.

$(x^2 - 4x + 4) = 2y - \dfrac{3}{2} + 4$ Add the square of half the coefficient of $-4x$ to both sides. $\left[\dfrac{1}{2}(-4)\right]^2 = \left[-2\right]^2 = 4$

$$-\dfrac{3}{2} + 4 = -\dfrac{3}{2} + \dfrac{8}{2} = \dfrac{5}{2}$$

$(x - 2)^2 = 2y + \dfrac{5}{2}$ Factor 2 out of the right side of the equation.

$(x - 2)^2 = 2\left(y + \dfrac{5}{4}\right)$ Now that we've completed the square, let's compare this equation to the standard form of a parabola.

$(x - h)^2 = 4p(y - k)$ $4p = 2$, $p = \dfrac{1}{2}$ Since p is positive and the x is squared, we know the parabola opens up.

$V: \left(2, -\dfrac{5}{4}\right)$, $F: \left(2, -\dfrac{5}{4} + \dfrac{1}{2}\right) = \left(2, -\dfrac{3}{4}\right)$, $D: y = -\dfrac{5}{4} + -\dfrac{1}{2} = -\dfrac{5}{4} - \dfrac{2}{4} = -\dfrac{7}{4}$

To find two other points on the graph easily, we'll create a perfect square on the right side of the equation by letting $y = \dfrac{3}{4}$. We chose $\dfrac{3}{4}$ to create a perfect square, but you could use a different number.

$(x - 2)^2 = 2\left(y + \dfrac{5}{4}\right)$ Substitute $\dfrac{3}{4}$ for y.

$(x - 2)^2 = 2\left(\dfrac{3}{4} + \dfrac{5}{4}\right)$ $\dfrac{3}{4} + \dfrac{5}{4} = \dfrac{8}{4} = 2$

$(x - 2)^2 = 2(2)$

$(x - 2)^2 = 4$ Take the square root of both sides of the equation.

$x - 2 = \pm\sqrt{4}$

$x - 2 = \pm 2$ Move the 2 to the right side of the equation.

$x = 2 \pm 2$ Separate the equation into two equations.

$x = 2 + 2 = 4, \ x = 2 - 2 = 0$

Two points on the parabola are $\left(4, \dfrac{3}{4}\right)$ and $\left(0, \dfrac{3}{4}\right)$.

The following is the graph of $2x^2 - 8x - 4y + 3 = 0$.

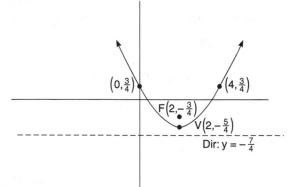

SELF-TEST 4

1. Find the value of p for the following parabola. In which direction does the graph open?

$x^2 + 6x + 4y + 5 = 0$

2. Find the vertex, directrix, and focus of the following parabola.

   $y^2 - y - x + 6 = 0$

3. Find the value of the vertex for the following parabola. In which direction does the graph open?

   $2x^2 - x = y + 1$

4. Graph the following parabola. Label the vertex, focus, directrix, and two other points on the parabola.

   $4y^2 - 23 = 16x + 12y$

ANSWERS

1. The x is squared, not the y, so we know this graph opens up or down. We have to find the value of p to determine which direction. To find p, we'll complete the square.

$x^2 + 6x + 4y + 5 = 0$	Move the $4y$ and the 5 to the other side of the equation.
$x^2 + 6x = -4y - 5$	Add the square of half the coefficient of $6x$ to both sides.
$x^2 + 6x + 9 = -4y - 5 + 9$	$\left[\frac{1}{2}(6)\right]^2 = [3]^2 = 9$
	Factor the left side of the equation.
$(x + 3)^2 = -4y + 4$	Factor the right side of the equation.
$(x + 3)^2 = -4(y - 1)$	Compare this equation to the standard form of a parabola.
$(x - h)^2 = 4p(y - k)$	The vertex is $(-3,1)$. $4p = -4$, $p = -1$. The x is squared and p is negative, so the parabola opens down.

2. $y^2 - y - x + 6 = 0$ Move the $-x$ and 6 to the other side of the equation.

   $y^2 - y = x - 6$ Add the square of half the coefficient of the middle term to both sides of the equation. $\left[\frac{1}{2}(-1)\right]^2 = \left[-\frac{1}{2}\right]^2 = \frac{1}{4}$.

   $y^2 - y + \frac{1}{4} = x - 6 + \frac{1}{4}$

   $\left(y - \frac{1}{2}\right)^2 = x - \frac{23}{4}$ Let's compare this equation to the standard form.

   $(y - k)^2 = 4p(x - h)$

   The vertex is $\left(\frac{23}{4}, \frac{1}{2}\right)$. $x - \frac{23}{4} = 1\left(x - \frac{23}{4}\right)$ so $4p = 1$ and $p = \frac{1}{4}$.

   Since p is positive and the y is squared, the parabola open to the right.

   The focus is $\left(\frac{23}{4} + \frac{1}{4}, \frac{1}{2}\right) = \left(\frac{24}{4}, \frac{1}{2}\right) = \left(6, \frac{1}{2}\right)$.

   The directrix is the line $x = \frac{11}{2}$, $\left(\frac{23}{4} - \frac{1}{4} = \frac{22}{4} = \frac{11}{2}\right)$.

3. $2x^2 - x = y + 1$ The coefficient of the square isn't 1, so we'll divide by 2 on both sides of the equation. Now add the square of half the coefficient of $-\frac{1}{2}x$ to both sides of the equation.

$$x^2 - \frac{1}{2}x = \frac{1}{2}y + \frac{1}{2} \qquad \left[\frac{1}{2}\left(-\frac{1}{2}\right)\right]^2 = \left[-\frac{1}{4}\right]^2 = \frac{1}{16}$$

$$\left(x - \frac{1}{2}x + \frac{1}{16}\right) = \frac{1}{2}y + \frac{1}{2} + \frac{1}{16} \qquad \text{Factor the left side of the equation.}$$

$$\left(x - \frac{1}{4}\right)^2 = \frac{1}{2}y + \frac{9}{16} \qquad \text{To find } p, \text{ factor } \frac{1}{2} \text{ from the right side of the equation.}$$

$$\left(x - \frac{1}{4}\right)^2 = \frac{1}{2}\left(y + \frac{9}{8}\right) \qquad \frac{9}{16} \div \frac{1}{2} = \frac{9}{16}\left(\frac{2}{1}\right) = \frac{9}{8},\ 4p = \frac{1}{2},\ p = \frac{\frac{1}{2}}{4} = \frac{1}{2}\left(\frac{1}{4}\right) = \frac{1}{8}$$

$$(x - h)^2 = 4p(y - k)$$

p is positive and x is squared; therefore, this parabola opens up.

The vertex is $\left(\frac{1}{4}, -\frac{9}{8}\right)$.

4. $4y^2 - 23 = 16x + 12y$ Put the terms containing a y on the left and all other terms on the right.

$4y^2 - 12y = 16x + 23$ Divide all terms by 4 to make the coefficient of the square 1.

$$y^2 - 3y = 4x + \frac{23}{4} \qquad \left[\frac{1}{2}(-3)\right]^2 = \left[-\frac{3}{2}\right]^2 = \frac{9}{4}$$

$$y^2 - 3y + \frac{9}{4} = 4x + \frac{23}{4} + \frac{9}{4} \qquad \frac{23}{4} + \frac{9}{4} = \frac{32}{4} = 8$$

$$\left(y - \frac{3}{2}\right)^2 = 4x + 8 \qquad \text{To find } p, \text{ factor 4 out of the right side of the equation.}$$

$$\left(y - \frac{3}{2}\right)^2 = 4(x + 2) \qquad 4p = 4,\ p = 1. \text{ Since the } y \text{ is squared and } p \text{ is positive, this parabola opens to the right.}$$

The vertex is $\left(-2, \frac{3}{2}\right)$.

The focus is $\left(-2 + 1, \frac{3}{2}\right) = \left(-1, \frac{3}{2}\right)$. The directrix is $x = -2 - 1 = -3$.

To find two other points on the parabola, we have to create a perfect square on the right side of the equation.

$$\left(y - \frac{3}{2}\right)^2 = 4(-1 + 2) \qquad \text{Let's begin by letting } x = -1.$$

$$\left(y - \frac{3}{2}\right)^2 = 4 \qquad \text{Take the square root of both sides of the equation.}$$

$$\sqrt{\left(y - \frac{3}{2}\right)^2} = \mp\sqrt{4}$$

$$y - \frac{3}{2} = \pm 2 \qquad \text{Separate the equation into two equations.}$$

$$y = \frac{3}{2} + 2 = \frac{3}{2} + \frac{4}{2} = \frac{7}{2} \quad y = \frac{3}{2} - \frac{4}{2} = -\frac{1}{2}$$ Two points of the points on the parabola are $\left(-1, \frac{7}{2}\right)$ and $\left(-1, -\frac{1}{2}\right)$.

The following figure is the graph of $4y^2 - 23 = 16x + 12y$.

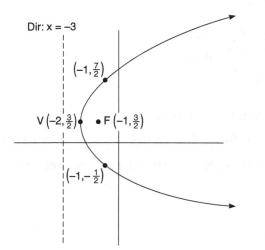

Ellipses

In this section we'll study another conic shape, the ellipse. We'll learn how to graph ellipses and write the equation of an ellipse.

An ellipse is the set of all points (x,y), the sum of whose distances from two distinct fixed points (the foci) is constant.

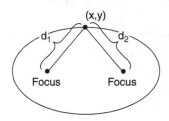

$d_1 + d_2$ is constant for every point on an ellipse.

A line drawn through the foci intersects the ellipse at two key points, called the vertices. The chord joining the vertices is the major axis, and its midpoint is the center of the ellipse. The chord perpendicular to the major axis is the minor axis of the ellipse. We'll label the center (h,k). We'll call the distance from the center to the vertex on the major axis a.

Therefore, the distance of the major axis is $2a$. The distance along the minor axis is $2b$. The major axis is always the longer axis. $0 < b < a$.

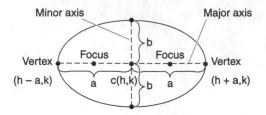

We'll call the distance from the center to the foci c.

The center of the ellipse is (h,k). The vertices are $(h + a,k)$ and $(h - a,k)$.

The foci are $(h + c,k)$ and $(h - c,k)$. By using the Pythagorean theorem, we can find the measure of the foci.

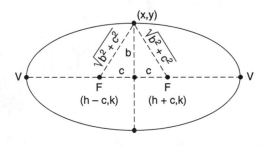

To derive the standard form of the equation of an ellipse, we'll use the fact that the sum of the distances from any point on the ellipse to the two foci is constant. At a vertex the constant sum is the length of the major axis $2a$.

$$\sqrt{[x - (h - c)]^2 + (y - k)^2} + \sqrt{[x - (h + c)]^2 + (y - k)^2} = 2a$$

Applying the Pythagorean theorem to the previous figure, we can see that $b^2 + c^2 = a^2$, which can be rewritten as $b^2 = a^2 - c^2$.

$$\sqrt{b^2 + c^2} + \sqrt{b^2 + c^2} = 2a$$

$$2\sqrt{b^2 + c^2} = 2a$$

$$\sqrt{b^2 + c^2} = a$$

The equation of the ellipse is:

$b^2(x - h)^2 + a^2(y - k)^2 = a^2b^2$ Divide both sides of the equation by a^2b^2.

$$\frac{(x - h)^2}{a^2} + \frac{(y - k)^2}{b^2} = 1$$

If the ellipse had a vertical major axis, we would have gotten the following equation.

$$\frac{(x-h)^2}{b^2} + \frac{(y-k)^2}{a^2} = 1$$

We will assume $a > b$ for all ellipses. If a^2 is located under the $(x - h)^2$ term, the major axis is horizontal. If a^2 is located under the $(y - k)^2$ term, the major axis is vertical.

Standard Equation of an Ellipse

The standard form of the equation of an ellipse, with center (h,k) and major and minor axes of length $2a$ and $2b$, where $0 < b < a$ is

$$\frac{(x-h)^2}{a^2} + \frac{(y-k)^2}{b^2} = 1 \qquad \text{Major axis is horizontal.}$$

$$\frac{(x-h)^2}{b^2} + \frac{(y-k)^2}{a^2} = 1 \qquad \text{Major axis is vertical.}$$

The foci lie on the major axis, c units from the center, with $c^2 = a^2 - b^2$. The vertices lie on the major axis, a units from the center.

Assume a^2 is the larger denominator. If a^2 is located under $(x - h)^2$, the major axis is horizontal. If a^2 is located under $(y - k)^2$, the major axis is vertical.

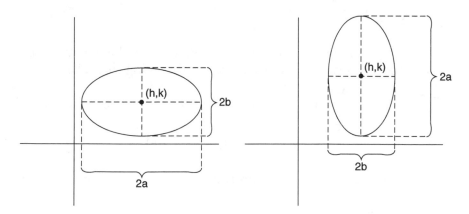

If the center is at the origin $(0,0)$, then the equation takes one of the following forms.

$$\frac{x^2}{a^2} + \frac{y^2}{b^2} = 1 \qquad \text{Major axis is horizontal. } a > b$$

$$\frac{x^2}{b^2} + \frac{y^2}{a^2} = 1 \qquad \text{Major axis is vertical. } a > b$$

Now that you know the standard equations, let's work on writing the equation of an ellipse given the graph of the ellipse.

Example 16:

Write the equation of the following ellipse.

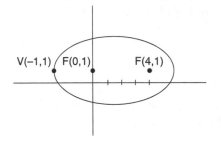

Solution:

We know the center is halfway between the foci. The distance from $(0,1)$ to $(4,1)$ is 4 units. This tells us that the center of the ellipse is $(2,1)$ and $c = 2$. The distance from the center to the vertex is a. For this ellipse, $a = 3$. Now that we know the values for a and c, we'll use the formula $b^2 = a^2 - c^2$ to find the value of b.

$b^2 = a^2 - c^2$	Take the square root of both sides of the equation.
$b = \sqrt{a^2 - c^2}$	Substitute 3 for a and 2 for c.
$b = \sqrt{9 - 4} = \sqrt{5}$	

We have all the information we need to write the equation of the ellipse. The major axis is horizontal, so we'll use the following formula.

$\dfrac{(x-h)^2}{a^2} + \dfrac{(y-k)^2}{b^2} = 1$	Substitute 2 for h, 1 for k, 3 for a, and $\sqrt{5}$ for b.
$\dfrac{(x-2)^2}{3^2} + \dfrac{(y-1)^2}{(\sqrt{5})^2} = 1$	Simplify $3^2 = 9$ and $\left(\sqrt{5}\right)^2 = 5$.
$\dfrac{(x-2)^2}{9} + \dfrac{(y-1)^2}{5} = 1$	This is the equation of the given ellipse.

Example 17:

Sketch the following ellipse: $\dfrac{x^2}{144} + \dfrac{y^2}{169} = 1$

Label the center, foci, and vertices.

Solution:

First we'll compare the given equation with the standard form of an ellipse.

$$\frac{(x-h)^2}{a^2} + \frac{(y-k)^2}{b^2} = 1$$

$$\frac{x^2}{144} + \frac{y^2}{169} = 1$$ This equation can also be written in the following form.

$$\frac{(x-0)^2}{144} + \frac{(y-0)^2}{169} = 0$$

The center (h,k) is $(0,0)$. The larger denominator, 169, is a^2, $a = 13$, $b^2 = 144$, $b = 12$. Because the larger denominator is under the y^2, the major axis is vertical. The major axis is $2a$ units, or $2(13) = 26$ units. The minor axis is $2b$ units, or $2(12) = 24$ units. Now that we know the measure of a and b, we'll use the formula $c^2 = a^2 - b^2$ to find the measure of c.

$c^2 = a^2 - b^2$ Substitute 13 for a and 12 for b.

$c^2 = 13^2 - 12^2$

$c^2 = 169 - 144$

$c^2 = 25$ Take the square root of both sides of the equation.

$c = 5$

Now that we know the measure of c, we can find the foci. The foci of an ellipse with a major axis that's vertical are $(h, c \pm k)$.

To find the foci, we remember that the foci are c units from the center on the major axis. The center is at $(0,0)$ and the major axis is vertical, so to find one of the foci, we move up 5 units from $(0,0)$ and get $(0,5)$. To find the other focus, we move 5 units down from $(0,0)$ and get $(0,-5)$. All we need now are the vertices and we're ready to graph the ellipse. The vertices of an ellipse with a major axis that's vertical are $(h, a \pm k)$. The vertices are a units from the center along the major axis. The center is at $(0,0)$ and a is 13, so the vertices are $(0,13)$ and $(0,-13)$. The graph follows.

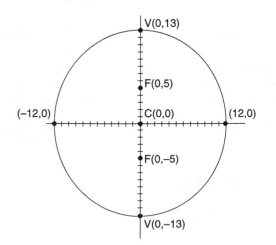

Example 18:

Find the center, foci, and vertices of the following ellipse.

$$\frac{(x-2)^2}{16} + \frac{(y+1)^2}{4} = 1$$

Solution:

Let's start by comparing the given equation to the standard form of an ellipse.

$$\frac{(x-h)^2}{a^2} + \frac{(y-k)^2}{b^2} = 1$$

$$\frac{(x-2)^2}{16} + \frac{(y+1)^2}{4} = 1$$

We can easily see the center is (2,–1). The larger denominator 16, which is always a^2, is below x, not y; therefore, the major axis is horizontal. $a = 4$ and $b^2 = 4$, so $b = 2$.

$$c^2 = a^2 - b^2 = 16 - 4 = 12$$

$$c = \sqrt{12} = \sqrt{4(3)} = 2\sqrt{3} \qquad h = 2, \ k = -1, \ a = 4, \ b = 2, \ c = 2\sqrt{3}$$

C: (2,–1)
V: $(h \pm a, k) = (2 \pm 4, -1) = (6, -1)$ and $(-2, -1)$
F: $(h \pm c, k) = (2 \pm 2\sqrt{3}, -1) = (2 + 2\sqrt{3}, -1)$ and $(2 - 2\sqrt{3}, -1)$

Example 19:

Write the equation of the ellipse with center at the origin, one focus at (3,0), and vertex at (–4,0).

Solution:

The change in the focus (3,0) and the vertex (–4,0) is a horizontal change. We know this because the x values are different while the y values are the same. This tells us that the major axis is horizontal. The distance from the center point (0,0) to a focus (3,0) is 3 units, so we know $c = 3$. The distance from the center point (0,0) to a vertex (–4,0) is 4 units, so we know $a = 4$. To find the value of b, we'll use the formula $b^2 = a^2 - c^2 = 4^2 - 3^2 = 16 - 9 = 7$, $b = \sqrt{7}$. All we have to do now is substitute these values into the standard form of an ellipse centered at the origin with a horizontal axis.

$$\frac{x^2}{a^2} + \frac{y^2}{b^2} = 1 \qquad \qquad \text{Substitute 4 for } a, \ \sqrt{7} \text{ for } b.$$

$$\frac{x^2}{16} + \frac{y^2}{7} = 1$$

By now you should be able to try a few on your own. Try the following problems, then check your answers with ours.

SELF-TEST 5

1. Draw the graph of the following ellipse. Label the center, foci, and vertices.
   $$\frac{x^2}{4} + \frac{y^2}{16} = 1$$

2. Draw the following ellipse. Label the center, foci, and vertices.
   $$\frac{(x-1)^2}{16} + \frac{(y+2)^2}{9} = 1$$

3. Write the equation of the ellipse with vertices (0,–8), (0,8), and foci (0,–4), (0,4).

4. Write the equation of the ellipse with foci (–4,–1), (–4,7) and a major axis of length 12.

5. Write the equation of the following ellipse.

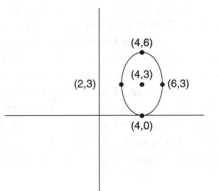

ANSWERS

1. $\frac{x^2}{4} + \frac{y^2}{16} = 1$ is of the form $\frac{(x-h)^2}{b^2} + \frac{(y-k)^2}{a^2} = 1$. We know this because the larger denominator, 16, is under the y^2. That tells us the major axis of the ellipse is vertical. a^2 is always the larger denominator, so we know $a^2 = 16$; therefore, $a = 4$. $b^2 = 4$ so $b = 2$. The length of the major axis is 2a units. The length of the major axis for the given ellipse is 2(4) = 8 units. The length of the minor axis is 2b units. The length of the minor axis for the given ellipse is 2(2) = 4 units. To find the measure of c, we'll use the formula $c^2 = a^2 - b^2$.

$c^2 = 4^2 - 2^2 = 16 - 4 = 12$, $c = \sqrt{12} = \sqrt{4(3)} = 2\sqrt{3}$. The center (h,k) is $(0,0)$. So $h = 0$, $k = 0$. The vertices of an ellipse with a vertical major axis are $(h, k \pm a)$. The vertices of the given ellipse are $(0,-4)$ and $(0,4)$. The foci of an ellipse with a vertical major axis are $(h, k \pm c)$. The foci of the given ellipse are $(0,-2\sqrt{3})$ and $(0,2\sqrt{3})$. All the points we have so far are on the minor axis. It would be helpful to know the two end points on the minor axis when we graph the ellipse. Because we know $b = 2$, two points on the minor axis are $(-2,0)$ and $(2,0)$. All we have to do now is draw all this information and we're done.

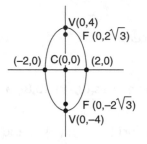

2. We know this ellipse is not centered at the origin. The center is $(1,-2)$, $h = 1$, $k = -2$. Because the larger denominator, 16, is located under the x^2, we know the major axis of the ellipse is horizontal and the minor axis is vertical. $a^2 = 16$, so $a = 4$. The length of the major axis is $2a$ units, which is $2(4) = 8$ units. $b^2 = 9$, so $b = 3$. The length of the minor axis is $2b$ units, which is $2(3) = 6$ units. Now that we know the values of a and b, we'll find the value of c by using the formula $c^2 = a^2 - b^2$. $c^2 = 16 - 9 = 7$. $c = \sqrt{7}$. The vertices of an ellipse with a horizontal major axis are $(h \pm a, k)$. The vertices of this ellipse are $(1 \pm 4, -2)$, which are $(-3,-2)$ and $(5,-2)$. The foci of an ellipse with a horizontal major axis are $(h \pm c, k)$. The foci of this ellipse are $(1 \pm \sqrt{7}, -2)$, which are $(1 - \sqrt{7}, -2)$ and $(1 + \sqrt{7}, -2)$. To find the end points of the minor axis, we'll add and subtract the measure of b to the y coordinate of the center point $(h, k \pm b)$. The end points of the minor axis of this ellipse are $(1,1)$ and $(1,-5)$. The following figure shows the graph of all this information.

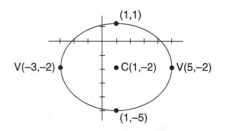

3. The vertices and foci form a vertical line; therefore, we know this ellipse has a major axis that is vertical. We'll use the formula $\dfrac{(x-h)^2}{b^2} + \dfrac{(y-k)^2}{a^2} = 1$ to write the equation of the ellipse. The distance from the vertices $(0,8)$ and $(0,-8)$ is the distance of the major axis, which in this case is 16 units. The major axis is $2a$ units, so a is 8. The center of the ellipse is a units from either of the vertices (or we could say the center is halfway between the vertices). The center of this ellipse is $(0,0)$, $h = 0$, $k = 0$. The foci $(0,-4)$ and $(0,4)$ are both

c units from the center $(0,0)$. The measure of c is 4. We'll use the formula $c^2 = a^2 - b^2$ to find the value of b.

$c^2 = a^2 - b^2$	Substitute 8 for a and 4 for c.
$4^2 = 8^2 - b^2$	Square the 4 and the 8.
$16 = 64 - b^2$	Subtract 64 on both sides of the equation.
$-48 = -b^2$	Multiply or divide both sides of the equation by -1.
$48 = b^2$	Take the square root of both sides of the equation.

$b = \sqrt{48} = \sqrt{16(3)} = 4\sqrt{3}$

Now that we know the values of a, b, h, and k, we're ready to substitute these values into the equation of an ellipse with a vertical major axis.

$\dfrac{(x-h)^2}{b^2} + \dfrac{(y-k)^2}{a^2} = 1$	Substitute $a = 8$, $b = 4\sqrt{3}$, $h = 0$, $k = 0$ into the equation.
$\dfrac{(x-0)^2}{(4\sqrt{3})^2} + \dfrac{(y-0)^2}{8^2} = 1$	Now simplify. $(4\sqrt{3})^2 = (4\sqrt{3})(4\sqrt{3}) = 16(3) = 48$
$\dfrac{x^2}{48} + \dfrac{y^2}{64} = 1$	

4. The fact that the foci have the same x coordinate tells us that the major axis is vertical. Therefore, we'll use the following formula to write the equation of the ellipse: $\dfrac{(x-h)^2}{b^2} + \dfrac{(y-k)^2}{a^2} = 1$. The halfway mark between the foci is the center of the ellipse. Halfway between $(-4,-1)$ and $(-4,7)$ is the center $(-4, 3)$. $h = -4$, $k = 3$. To find the measure of c, we find the distance between one of the foci and the center point. $c = 4$. $2a$ is the length of the major axis. Since we know the major axis is 12, $a = 6$. Now all we have to do is find the measure of b by using the formula $c^2 = a^2 - b^2$.

$c^2 = a^2 - b^2$	Substitute 4 for c and 6 for a.
$4^2 = 6^2 - b^2$	Square 4 and 6.
$16 = 36 - b^2$	Subtract 36 from both sides of the equation.
$-20 = -b^2$	Multiply or divide both sides of the equation by -1.
$20 = b^2$	Take the square root of both sides of the equation.

$b = \sqrt{20} = \sqrt{4(5)} = 2\sqrt{5}$

To write the equation of the ellipse, substitute 6 for a, $2\sqrt{5}$ for b, -4 for h, and 3 for k into the standard form of an ellipse with a vertical major axis.

$$\frac{(x-h)^2}{b^2} + \frac{(y-k)^2}{a^2} = 1$$

$$\frac{(x--4)^2}{(2\sqrt{5})^2} + \frac{(y-3)^2}{6^2} = 1$$

$$\frac{(x+4)^2}{20} + \frac{(y-3)^2}{36} = 1$$

5. We can see by looking at the graph that the center point is (4,3). $h = 4$, $k = 3$. The longer axis, which is the major axis, is vertical; therefore, we will use the formula $\dfrac{(x-h)^2}{b^2} + \dfrac{(y-k)^2}{a^2} = 1$ to write the equation of the ellipse. The distance from a vertex to the center is 3 units, so $a = 3$. The distance from an end point on the minor axis to the center is 2 units, so $b = 2$. Using this information, we find the equation of the ellipse is:

$$\frac{(x-h)^2}{b^2} + \frac{(y-k)^2}{a^2} = 1$$

$$\frac{(x-4)^2}{2^2} + \frac{(y-3)^2}{3^2} = 1$$

$$\frac{(x-4)^2}{4} + \frac{(y-3)^2}{9} = 1$$

Now that you know how to write the equation of an ellipse in standard form and how to sketch an ellipse, we're going to do the same thing we did with circles and parabolas. We're going to use completing the square on the general form of an ellipse to convert the ellipse to standard form. Once the ellipse is in standard form, it's very easy to see the coordinates of the center point and the length of the major and minor axes.

Example 20:

Convert the following ellipse to standard form. Find the center and foci.

$4x^2 - 16x + 9y^2 + 18y = 11$

Solution:

An ellipse, unlike a circle, can have denominators other than 1. A circle doesn't show any denominators; an ellipse does. Unlike a circle, an ellipse is always equal to 1. To complete the square, the leading coefficients should be 1, but in the given equation they are 4 and 9. To change this, we'll begin by factoring out a 4 from the terms that have an x and factoring out a 9 from the terms that have a y.

$4x^2 - 16x + 9y^2 + 18y = 11$	Factor the 4 and the 9 out of the x and y terms.
$4(x^2 - 4x) + 9(y^2 + 2y) = 11$	Add the square of half the coefficients of the middle terms to both sides of the equation. Remember the constant you add inside the $(x^2 - 4x)$ is multiplied by 4 and the constant you add inside the $(y^2 + 2y)$ is multiplied by 9 when adding them to the right side of the equation.

$4(x^2 - 4x + 4) + 9(y^2 + 2y + 1) = 11 + 16 + 9$

We didn't have to do this with circles and parabolas because the equations of circles and parabolas aren't written as fractions. To find the square of half the coefficient of the x term, we have $\left[\frac{1}{2}(-4)\right]^2 = [-2]^2 = 4$. The 4 is added inside $4(x^2 - 4x)$, so we add 16 to the right side of the equation. To find the square of half the coefficient of the y term, we have $\left[\frac{1}{2}(2)\right]^2 = [1]^2 = 1$. The 1 is added inside $9(y^2 + 2y)$, so we add 9 to the right side of the equation.

$4(x^2 - 4x + 4) + 9(y^2 + 2y + 1) = 11 + 16 + 9$ Now factor and add the constants.

$4(x - 2)^2 + 9(y + 1)^2 = 36$

The standard form of the equation of an ellipse $\dfrac{(x - h)^2}{a^2} + \dfrac{(y - k)^2}{b^2} = 1$ looks like this. In order to write this equation in this form, we have to get rid of the 4 and the 9 and the 36 must be a 1. To accomplish this we'll divide the entire equation by 36.

$\dfrac{4(x - 2)^2}{36} + \dfrac{9(y + 1)^2}{36} = \dfrac{36}{36}$ Reduce the common factors.

$\dfrac{(x - 2)^2}{9} + \dfrac{(y + 1)^2}{4} = 1$ Finally the ellipse is in standard form.

The center is $(2, -1)$, $h = 2$, $k = -1$. The major axis is horizontal. $a = 3$, $b = 2$, the foci are c units from the center point. To find c we'll use the formula $c^2 = a^2 - b^2$.

$c^2 = a^2 - b^2$ Substitute 3 for a and 2 for b.

$c^2 = 3^2 - 2^2$ Square the 3 and the 2.

$c^2 = 9 - 4$

$c^2 = 5$ Take the square root of both sides of the equation.

$c = \pm\sqrt{5}$ The foci of an ellipse whose major axis is horizontal are $(h \pm c, k)$.

                    The foci of this ellipse are $(2 \pm \sqrt{5}, -1)$, or $(2 - \sqrt{5}, -1)$, $(2 + \sqrt{5}, -1)$.

Example 21:

Write the following ellipse in standard form. Find the center point, the vertices, and the foci.

$4x^2 + 9y^2 = 36$

Solution:

We don't have to complete the square on this equation because there isn't an x to the first power term or a y to the first-power term. We'll begin by dividing every term of the equation by 36 to get a 1 on the right side of the equation.

We know that this ellipse is centered at the origin because it's of the form $\frac{x^2}{a^2} + \frac{y^2}{b^2} = 1$. Therefore, we know $h = 0$ and $k = 0$. The constant on the right side of the equation has to be 1, not 36, so we'll divide both sides of the equation by 36.

$\frac{4x^2}{36} + \frac{9y^2}{36} = \frac{36}{36}$	Reduce the common factors.
$\frac{x^2}{9} + \frac{y^2}{4} = 1$	The value of a is 3 and b is 2. To find c we'll use the formula $c^2 = a^2 - b^2$.
$c^2 = a^2 - b^2$	Substitute 3 for a and 2 for b.
$c^2 = 3^2 - 2^2$	Square the 3 and the 2.
$c^2 = 5$	Take the square root of both sides of the equation.
$c = \pm\sqrt{5}$	The center point is $(0,0)$. The foci are $(h \pm c, k)$, which are $(-\sqrt{5}, 0)$, $(\sqrt{5}, 0)$. The vertices are $(h \pm a, k)$, which are $(-3, 0)$ and $(3, 0)$.

Example 22:

Convert the following equation to standard form and find the center point, the vertices, and the foci.

$$25x^2 + y^2 - 12y = -11$$

Solution:

This ellipse doesn't have an x to the first power term; therefore, we can't complete the square on the x term. We'll complete the square on the y terms.

$25x^2 + y^2 - 12y = -11$	$\left[\frac{1}{2}(-12)\right]^2 = [-6]^2 = 36$
$25x^2 + y^2 - 12y + 36 = -11 + 36$	Factor and combine the constants.
$25x^2 + (y - 6)^2 = 25$	Divide by 25.
$\frac{25x^2}{25} + \frac{(y - 6)^2}{25} = \frac{25}{25}$	Reduce the common factors.
$x^2 + \frac{(y - 6)^2}{25} = 1$	The center point is $(0,6)$, $h = 0$, $k = 6$. The major axis is vertical. $a = 5$, $b = 1$.

$$c^2 = a^2 - b^2$$

$$c^2 = 25 - 1$$

$$c = \sqrt{24} = \sqrt{4(6)} = 2\sqrt{6}$$ The foci are $(h, k \pm c)$ or $(0, 6 - 2\sqrt{6})$, $(0, 6 + 2\sqrt{6})$.

The vertices are $(h, k \pm a)$ or $(0, 11)$, $(0, 1)$.

Example 23:

Find the center point, vertices, and foci of the following ellipse.

$4x^2 + 24x + 13y^2 - 26y = 13$

Solution:

The leading coefficients aren't 1, they're 4 and 13, so we'll start by factoring out 4 from the terms with an *x* and 13 from the terms that have a *y*.

$4(x^2 + 6x) + 13(y^2 - 2y) = 13$ We'll add the square of half the coefficients of the first power terms to each side of the equation.

$4(x^2 + 6x + 9) + 13(y^2 - 2y + 1) = 3 + 36 + 13$ Factor, then add the constants.

$4(x + 3)^2 + 13(y - 1)^2 = 52$ The equation of an ellipse is set equal to 1, but this one is equal to 52, so we'll divide the equation by 52.

$$\frac{4(x + 3)^2}{52} + \frac{13(y - 1)^2}{52} = \frac{52}{52}$$ Reduce the common factors.

$$\frac{(x + 3)^2}{13} + \frac{(y - 1)^2}{4} = 1$$ The center point is $(-3, 1)$, $h = -3$, $k = 1$.

The value of *a* is $\sqrt{13}$. The value of *b* is 2.

This ellipse has a horizontal major axis.

Let's find the value of *c*.

$$c^2 = a^2 - b^2$$

$$c^2 = (\sqrt{13})^2 - 2^2$$

$$c^2 = 13 - 4 = 9$$

$$c = 3$$ The vertices are $\left(-3 - \sqrt{13}, 1\right)$, $\left(-3 + \sqrt{13}, 1\right)$

The foci are $(0, 1)$, $(-6, 1)$

1. Find the center point and the values of a, b, and c for the following ellipse. Is the major axis horizontal or vertical?

$$9x^2 + 3y^2 = 9$$

2. Find the center point, foci, and vertices of the following ellipse.

$$3(x + 2)^2 + 4(y - 1)^2 = 192$$

3. Find the center point, foci, and vertices of the following ellipse.

$$9x^2 + 4y^2 - 18x + 16y - 11 = 0$$

4. Write the following ellipse in standard form. Graph the ellipse. Label the center point and four other points on the circumference of the ellipse.

$$16x^2 + 9y^2 - 128x + 54y + 193 = 0$$

ANSWERS

1. $9x^2 + 3y^2 = 9$

The right side of the equation should be 1, so we'll begin by dividing all terms by 9.

$$\frac{9x^2}{9} + \frac{3y^2}{9} = \frac{9}{9}$$

Reduce the common factors.

$$\frac{x^2}{1} + \frac{y^2}{3} = 1$$

The larger denominator is under the y^2, so the major axis is vertical. $a = \sqrt{3}$, $b = 1$, $c^2 = a^2 - b^2$. The center point is (0,0).

$$c = \sqrt{3 - 1} = \sqrt{2}$$

2. $3(x + 2)^2 + 4(y - 1)^2 = 192$

To convert this to standard form, we'll divide by 192.

$$\frac{3(x + 2)^2}{192} + \frac{4(y - 1)^2}{192} = \frac{192}{192}$$

Reduce the common factors.

$$\frac{(x + 2)^2}{64} + \frac{(y - 1)^2}{48} = 1$$

The major axis is horizontal, $a = 8$.

$$b = \sqrt{48} = \sqrt{16(3)} = 4\sqrt{3}$$

$$c^2 = a^2 - b^2$$

Substitute 8 for a and $4\sqrt{3}$ for b.

$$c^2 = 8^2 - (4\sqrt{3})^2 = 64 - 48 = 16$$

$$c = 4$$

The center point is (–2,1). The foci are (–2 ± 4,1), which is (–6,1) and (2,1).

The vertices are (–2 ± 8,1), which is (–10,1), and (6,1).

3. To convert this ellipse from general form to standard form, we have to use the process of completing the square. We'll factor 9 out of the terms that contain an x and factor 4 out of the terms that contain a y.

$9x^2 + 4y^2 - 18x + 16y - 11 = 0$

$9(x^2 - 2x) + 4(y^2 + 4y) = 11$ Add the square of half the coefficients of the first power terms to both sides of the equation.

$9(x^2 - 2x + 1) + 4(y^2 + 4y + 4) = 11 + 9 + 16$ Factor and add the constants.

$9(x - 1)^2 + 4(y + 2)^2 = 36$ Divide all terms by 36.

$\dfrac{9(x - 1)^2}{36} + \dfrac{4(y + 2)^2}{9} = 1$ Reduce the common factors.

$\dfrac{(x - 1)^2}{4} + \dfrac{(y + 2)^2}{9} = 1$ The center point is $(1,-2)$, $a = 3$, $b = 2$.

The major axis is vertical. $c = \sqrt{5}$.

The foci are $(1,-2 - \sqrt{5})$ and $(1,-2 + \sqrt{5})$

The vertices are $(1,-5)$ and $(1,1)$.

4. $16x^2 + 9y^2 - 128x + 54y + 193 = 0$ We'll complete the square to convert this ellipse to standard form.

$16(x^2 - 8x) + 9(y^2 + 6y) = -193$

$16(x^2 - 8x + 16) + 9(y^2 + 6y + 9) = -193 + 256 + 81$

$16(x - 4)^2 + 9(y + 3)^2 = 144$

$\dfrac{16(x - 4)^2}{144} + \dfrac{9(y + 3)^2}{144} = \dfrac{144}{144}$

$\dfrac{(x - 4)^2}{9} + \dfrac{(y + 3)^2}{16} = 1$

The center point is $(4,-3)$. $a = 4$, $b = 3$. The major axis is vertical. To find two points on the circumference of the ellipse, we have to move 4 units up and 4 units down from the center point. This gives us the points $(4,-7)$ and $(4,1)$. To find two more points on the circumference of the ellipse, we have to move to the right 3 units and to the left 3 units from the center point. This gives us the points $(1,-3)$ and $(7,-3)$. The graph follows.

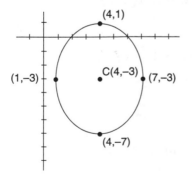

Hyperbolas

So far we've studied circles, ellipses, and parabolas. The last type of conic section we'll study is the hyperbola. We'll follow the same format as the rest of the chapter. Let's begin by looking at hyperbolas in standard form and finish with general form.

A hyperbola is the set of all points (x,y), the difference of whose distances from two distinct fixed points called the foci is constant.

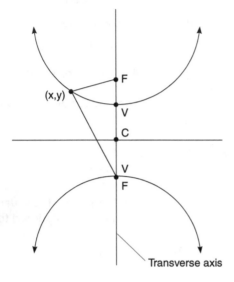

This is very similar to the definition of an ellipse, except the definition of an ellipse says "the sum," not "the difference." A hyperbola has two disconnected parts called the branches of the hyperbola. It also has two vertices, a center point, and two foci. The line segment connecting the vertices is called the transverse axis. The midpoint of the transverse axis is the center point of the hyperbola.

The standard form of the equation of a hyperbola centered at the origin is

$$\frac{x^2}{a^2} - \frac{y^2}{b^2} = 1$$ (transverse axis is horizontal) Asymptotes: $y = \pm\frac{b}{a}x$

Foci: $(\pm c, 0)$ Vertices: $(\pm a, 0)$

or

$$\frac{y^2}{a^2} - \frac{x^2}{b^2} = 1$$ (transverse axis is vertical) Asymptotes: $y = \pm\frac{a}{b}x$

Foci: $(0, \pm c)$ Vertices: $(0, \pm a)$

$a \neq 0$, $b \neq 0$. The vertices and foci are, respectively, a and c units from the center and $b^2 = c^2 - a^2$. We'll call a the distance from the center point to a vertex. We'll call c the distance from the center point to a focus.

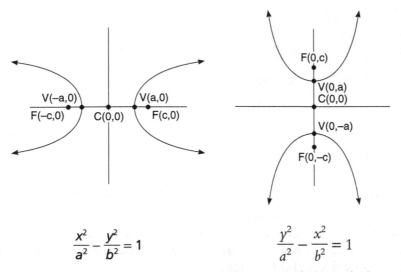

$$\frac{x^2}{a^2} - \frac{y^2}{b^2} = 1$$

$$\frac{y^2}{a^2} - \frac{x^2}{b^2} = 1$$

Lines called asymptotes can help us graph hyperbolas. The asymptotes pass through the corners of a rectangle of dimensions $2a$ by $2b$ and intersect at the center of the hyperbola.

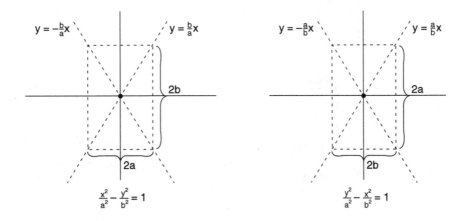

Example 24:

Find the vertices, foci, and center point of the following hyperbola. Sketch the graph.

$$\frac{y^2}{9} - \frac{x^2}{1} = 1$$

Solution:

Since the y is squared, the transverse axis is vertical. $a = 3$ and $b = 1$. To find c, we'll use the formula $b^2 = c^2 - a^2$.

$$\frac{y^2}{a^2} - \frac{x^2}{b^2} = 1 \qquad\qquad \frac{y^2}{9} - \frac{x^2}{1} = 1$$

$b^2 = c^2 - a^2$ — Substitute 3 for a and 1 for b.

$1^2 = c^2 - 3^2$ — Square 3 and 1.

$1 = c^2 - 9$ — Add 9 to both sides of the equation.

$10 = c^2$ — Take the square root of both sides of the equation.

$c = \sqrt{10} \approx 3.16$

The asymptotes are $y = \pm\dfrac{a}{b}x = \pm 3x$, $y = -3x$, $y = 3x$. The foci are $(0,-c)$ and $(0,c)$, $\left(0,-\sqrt{10}\right)$, $\left(0,\sqrt{10}\right)$. The vertices are $(0,-a)$ and $(0,a)$, $(0,-3)$, $(0,3)$. The center point is $(0,0)$. The rectangle the asymptotes pass through is $2a$ by $2b$, which is 6 by 2. Since the transverse axis is vertical, the 6 is the vertical dimension of the box and the 2 is horizontal dimension.

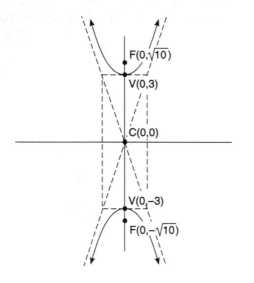

Example 25:

Graph the following hyperbola. Label the foci, center point, vertices, and asymptotes.

$$\frac{x^2}{7} - \frac{y^2}{9} = 1$$

Solution:

x^2 is first, so the transverse axis is horizontal. $a^2 = 7$ so $a = \sqrt{7}$, $b = 3$. We'll find c by using the formula $b^2 = c^2 - a^2$.

$b^2 = c^2 - a^2$ Substitute $\sqrt{7}$ for a and 3 for b.

$3^2 = c^2 - (\sqrt{7})^2$

$9 = c^2 - 7$ Add 7 to both sides of the equation.

$16 = c^2$ Take the square root of both sides of the equation.

$c = 4$

The foci are $(-4,0)$ and $(4,0)$. The vertices are $\left(-\sqrt{7},0\right)$ and $\left(\sqrt{7},0\right)$.

The center point is $(0,0)$. The asymptotes are $y = \pm \dfrac{b}{a}x$, $y = \pm \dfrac{3}{\sqrt{7}x} = \pm \dfrac{3\sqrt{7}}{7}x$.

The box is $2a$ by $2b$, or $2\sqrt{7}$ by 6. The $2\sqrt{7}$ is horizontal and the 6 is vertical.

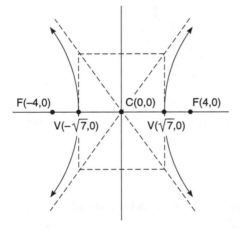

The following example is a hyperbola, but it's not in standard form. We can see this because it's not equal to 1. The equation of a hyperbola in standard form is set equal to 1.

Example 26:

Graph the hyperbola $4x^2 - y^2 = 16$.

Solution:

The standard form of the equation of a hyperbola is equal to 1, so we'll begin by dividing every term of the equation by 16.

$\dfrac{4x^2}{16} - \dfrac{y^2}{16} = \dfrac{16}{16}$ Reduce the common factors.

$\dfrac{x^2}{4} - \dfrac{y^2}{16} = 1$ Now that this equation is in standard form, the rest is easy.

Because x^2 appears first in the equation, the transverse axis is horizontal. $a = 2$, $b = 4$, $c = \sqrt{a^2 + b^2} = \sqrt{4 + 16} = \sqrt{20} \approx 4.47$. The foci are $\left(\pm\sqrt{20}, 0\right)$, the vertices are $(\pm 2, 0)$. The center point is $(0,0)$. The rectangle is 4 by 8. The asymptotes are $y = \pm\dfrac{b}{a}x$, $y = \pm\dfrac{4}{2}x$, $y = -2x$ and $y = 2x$.

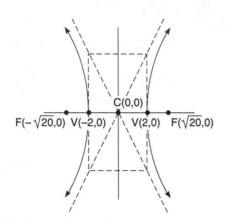

Example 27:

Find the vertices, foci, and asymptotes of the following hyperbola.

$3y^2 - 5x^2 = 15$

Solution:

The y^2 is first, so the transverse axis is vertical and the standard form is $\dfrac{y^2}{a^2} - \dfrac{x^2}{b^2} = 1$.

(The given example is equal to 15, but the standard form is equal to 1. The first thing we have to do is divide every term of the given equation: $3y^2 - 5x^2 = 15$ by 15.)

$\dfrac{3y^2}{15} - \dfrac{5x^2}{15} = \dfrac{15}{15}$	Reduce the common factors.
$\dfrac{y^2}{5} - \dfrac{x^2}{3} = 1$	$a^2 = 5$, $a = \sqrt{5}$, $b^2 = 3$, $b = \sqrt{3}$ Now let's find c.
$b^2 = c^2 - a^2$	Substitute 3 for b^2 and 5 for a^2.
$3 = c^2 - 5$	Add 5 to both sides of the equation then take its
$8 = c^2$	square root.

$c = \sqrt{8} = 2\sqrt{2} \approx 2.83$

The asymptotes are $y = \pm\dfrac{a}{b}x = \pm\dfrac{\sqrt{5}}{\sqrt{3}}x$. The vertices are $\left(0, \pm\sqrt{5}\right)$, and the foci are $\left(0, \pm 2\sqrt{2}\right)$.

In every problem we've done so far, we've given you the equation of the hyperbola and asked you to find the vertices, foci, and/or center point. In most cases we've also asked you to sketch the graph. Now it's time to mix things up a little. In the next few problems we'll give you the graph of a hyperbola or some information about the hyperbola and ask you to give us the equation.

Example 28:

Find the equation of the hyperbola centered at the origin with focus (0,–6) and vertex (0,4). Find the equations of the asymptotes of the hyperbola.

Solution:

Notice the focus and vertex are on a vertical transverse axis and the center point is at (0,0). This information tells us that the standard form for this hyperbola is $\frac{y^2}{a^2} - \frac{x^2}{b^2} = 1$. The focus is 6 units below the center point; therefore, $c = 6$. The vertex is 4 units above the center point; therefore, $a = 4$. In order to write the equation of the hyperbola, we have to find the value of b.

$$b^2 = c^2 - a^2$$
$$b^2 = 6^2 - 4^2$$
$$b^2 = 36 - 16$$
$$b = \sqrt{20} = \sqrt{4(5)} = 2\sqrt{5} \approx 4.47$$

Substitute 6 for c and 4 for a, then take the square root of both sides of the equation.

Substitute the values for a and b into the standard equation $\frac{y^2}{a^2} - \frac{x^2}{b^2} = 1$. $\frac{y^2}{16} - \frac{x^2}{20} = 1$ is the equation of the hyperbola with the given specifications. The asymptotes are $y = \pm\frac{a}{b}x = \pm\frac{4}{2\sqrt{5}}x = \pm\frac{2}{\sqrt{5}}x$.

All of the examples we've worked on so far have had center points at the origin. Let's work on a couple of problems whose center points are not at the origin. The following are the standard formulas of hyperbolas whose center points don't have to be located at the origin. The center point is (h,k).

$$\frac{(x-h)^2}{a^2} - \frac{(y-k)^2}{b^2} = 1$$

(Horizontal transverse axis)

C: (h,k), V: $(h \pm a,k)$, F: $(h \pm c,k)$

Asymptotes: $y - k = \pm\frac{b}{a}(x - h)$

or

$$\frac{(y-k)^2}{a^2} - \frac{(x-h)^2}{b^2} = 1$$

(Vertical transverse axis)

C: (h,k), V: $(h,k \pm a)$, F: $(h,k \pm c)$

Asymptotes: $y - k = \pm\frac{a}{b}(x - h)$

If you keep in mind that the box is $2a$ by $2b$, you'll have no trouble drawing the asymptotes.

Example 29:

Graph the following hyperbola. Label the vertices, foci, asymptotes, and center point.

$$\frac{(x+1)^2}{144} - \frac{(y-4)^2}{25} = 1$$

Solution:

The x is first, so we can assume the transverse axis is horizontal. The standard form is $\dfrac{(x-h)^2}{a^2} - \dfrac{(y-k)^2}{b^2} = 1$. The center point is $(-1,4)$, a $= 12$, and $b = 5$. $c = \sqrt{a^2 + b^2} = \sqrt{144 + 25} = \sqrt{169} = 13$. The foci are c units from the center point, so the foci are $(-1 \pm 13, 4)$, which are $(-14,4)$ and $(12,4)$. The vertices are a units from the center point, so the vertices are $(-1 \pm 12, 4)$, which are $(-13,4)$ and $(11,4)$. The box the asymptotes pass through is $2a$ by $2b$, which is 24 by 10. The asymptotes are:

$$y - k = \pm\frac{b}{a}(x-h)$$

$$y - 4 = \frac{5}{12}(x+1) \qquad \text{and} \qquad y - 4 = -\frac{5}{12}(x+1)$$

$$y - 4 = \frac{5}{12}x + \frac{5}{12} \qquad\qquad y - 4 = -\frac{5}{12}x - \frac{5}{12}$$

$$y = \frac{5}{12}x + \frac{5}{12} + 4 \qquad\qquad y = -\frac{5}{12}x - \frac{5}{12} + 4$$

$$y = \frac{5}{12}x + \frac{5}{12} + \frac{48}{12} \qquad\qquad y = -\frac{5}{12}x - \frac{5}{12} + \frac{48}{12}$$

$$y = \frac{5}{12}x + \frac{53}{12} \qquad\qquad y = -\frac{5}{12}x + \frac{43}{12}$$

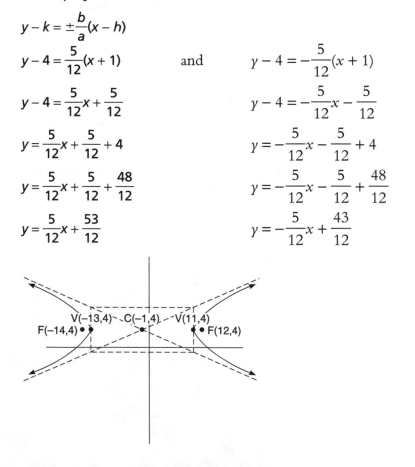

Example 30:

Use the following graph to write the equation of the hyperbola.

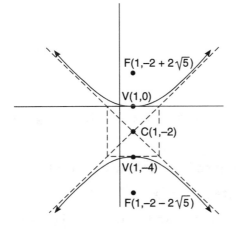

Solution:

The center point is (1,−2), so we know $h = 1$ and $k = −2$. The transverse axis is vertical, so we know the standard equation of this hyperbola is $\dfrac{(y - k)^2}{a^2} - \dfrac{(x - h)^2}{b^2} = 1$. To find the values of a and b, remember the vertices are a units from the center point. The vertices (1,0) and (1,−4) are 2 units from the center point, so a is 2. The foci are c units from the center point. The foci $(1,−2 + 2\sqrt{5})$ and $(1,−2 − 2\sqrt{5})$ are $2\sqrt{5}$ units from the center point, so $c = 2\sqrt{5}$. To find the value of b, we'll use a variation of the formula $c^2 = a^2 + b^2$.

$b^2 = c^2 - a^2$ Isolate the b^2.

$b = \sqrt{(2\sqrt{5})^2 - (2)^2} = \sqrt{20 - 4} = \sqrt{16} = 4$ Take the square root of both sides of the equation.

Because we know $h = 1$, $k = −2$, $a = 2$, and $b = 4$, we can substitute these values into the standard equation and find the equation of the graph of the given hyperbola.

$$\dfrac{(y + 2)^2}{4} - \dfrac{(x - 1)^2}{16} = 1$$

Now that we've worked through the last seven examples with you, it's your turn to try a few problems.

SELF-TEST 7

1. Find the center point, vertices, and foci of the following hyperbola. Graph the hyperbola.

$$\frac{y^2}{1} - \frac{x^2}{9} = 1$$

2. Graph the following hyperbola. Label the center point, vertices, and asymptotes.

$$9y^2 - 36x^2 = 4$$

3. Find the coordinates of the vertices and foci of the following hyperbola.

$$4x^2 - y^2 = 4$$

4. Write the equation of the specified hyperbola with its center at the origin.

Foci: (−5,0) and (5,0). Asymptotes: $y = \pm\frac{3}{4}x$

5. Write the equation of the hyperbola with foci (0,−5) and (0,5); vertex at (0,3).

6. Graph the following hyperbola. Label the center point and vertices.

$$\frac{(x-3)^2}{4} - \frac{(y+2)^2}{9} = 1$$

7. Write the equation of the following hyperbola.

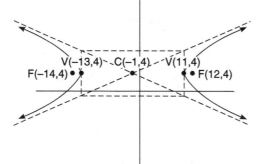

ANSWERS

1. $\dfrac{y^2}{1} - \dfrac{x^2}{9} = 1$

Because the y^2 appears first, the transverse axis is vertical.

The center point is (0,0). $a = 1$, $b = 3$

$c = \sqrt{a^2 + b^2} = \sqrt{1+9} = \sqrt{10} \approx 3.16$

The vertices are (0,–1) and (0,1). The foci are $(0,-\sqrt{10})$ and $(0,\sqrt{10})$.

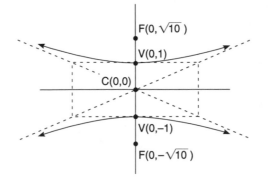

2. $9y^2 - 36x^2 = 4$

The equation of a hyperbola in standard form should be equal to 1, not 4, so we'll begin by dividing every term of the equation by 4.

$\dfrac{9y^2}{4} - \dfrac{36x^2}{4} = \dfrac{4}{4}$ Now the equation will be equal to 1, but the coefficients of the square terms are not 1. We'll have to rewrite the equation so the coefficients of the square terms are 1. In order for us to do this, we must realize that multiplication by $\dfrac{9}{4}$ is the same as division by $\dfrac{4}{9}$ and that multiplication by $\dfrac{9}{1}$ is the same as division by $\dfrac{1}{9}$.

$\dfrac{9y^2}{4} - \dfrac{9x^2}{1} = 1$

$\dfrac{y^2}{\frac{4}{9}} - \dfrac{x^2}{\frac{1}{9}} = 1$ $a^2 = \dfrac{4}{9},\ a = \dfrac{2}{3}\ b^2 = \dfrac{1}{9},\ b = \dfrac{1}{3}$

$c = \sqrt{a^2 + b^2} = \sqrt{\dfrac{4}{9} + \dfrac{1}{9}} = \sqrt{\dfrac{5}{9}} = \dfrac{\sqrt{5}}{3}$

The center point is (0,0). The transverse axis is vertical. The vertices are $\left(0, -\dfrac{2}{3}\right)$ and

$\left(0, \dfrac{2}{3}\right)$. The asymptotes are $y = \pm\dfrac{a}{b}x$, $y = \pm\dfrac{\frac{2}{3}}{\frac{1}{3}}x$, which is $y = -2x$ and

$y = 2x$.

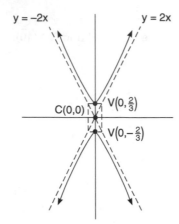

3. $4x^2 - y^2 = 4$

We'll begin by dividing all the terms of the equation by 4 to set the equation equal to 1.

$$\frac{4x^2}{4} - \frac{y^2}{4} = \frac{4}{4}$$

$$\frac{x^2}{1} - \frac{y^2}{4} = 1 \qquad a = 1, b = 2, c = \sqrt{1+4} = \sqrt{5} \approx 2.24$$

The transverse axis is horizontal. The center point is (0, 0). The vertices are (–1,0) and (1,0). The foci are $(-\sqrt{5},0)$ and $(\sqrt{5},0)$.

4. The halfway point between the foci is the center point. The halfway point between (–5,0) and (5,0) is (0,0). The center point is (0, 0). The foci form a horizontal line; therefore, the transverse axis is horizontal and the standard form of the equation of the hyperbola is $\frac{x^2}{a^2} - \frac{y^2}{b^2} = 1$. We also know the equation of the asymptotes is $y = \pm\frac{b}{a}x$. The given equation is $y = \pm\frac{3}{4}x$, so $a = 4$ and $b = 3$. The equation of the hyperbola is $\frac{x^2}{16} - \frac{y^2}{9} = 1$.

5. This problem is similar to number 4 except the transverse axis is vertical. The halfway point between the foci is (0,0), which is the center point of the hyperbola. The distance from the center point is c units, $c = 5$. The distance from the center point to the vertex (0,3) is a units, $a = 3$. $b = \sqrt{c^2 - a^2} = \sqrt{5^2 - 3^2} = \sqrt{25-9} = \sqrt{16} = 4$. The equation of the hyperbola is $\frac{y^2}{9} - \frac{x^2}{16} = 1$.

6. $\frac{(x-3)^2}{4} - \frac{(y+2)^2}{9} = 1 \quad C: (3,-2), a = 2, b = 3, c = \sqrt{4+9} = \sqrt{13} \approx 3.61$

The transverse axis is horizontal.

$V: (1,-2)$ and $(5,-2)$

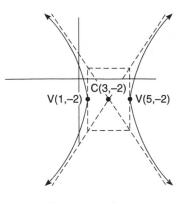

7. $C: (-1, 4)$, $a = 12$, $b = 5$.

$$\frac{(x + 1)^2}{144} - \frac{(y - 4)^2}{25} = 1$$

Hyperbolas—General Form

Now that you have the hang of hyperbolas in standard form, let's work on hyperbolas in general form. To derive the equations of hyperbolas in general form, we'll expand the equations of hyperbolas in standard form.

$$\frac{(x - h)^2}{a^2} - \frac{(y - k)^2}{b^2} = 1$$ Multiply all terms by a^2b^2.

$$\frac{a^2b^2(x - h)^2}{a^2} - \frac{a^2b^2(y - k)^2}{b^2} = a^2b^2(1)$$ Reduce the common factors.

$$b^2(x - h)^2 - a^2(y - k)^2 = a^2b^2$$

or

$$\frac{(y - k)^2}{a^2} - \frac{(x - h)^2}{b^2} = 1$$ Multiply all terms by a^2b^2.

$$\frac{a^2b^2(y - k)^2}{a^2} - \frac{a^2b^2(x - h)^2}{b^2} = a^2b^2(1)$$ Reduce the common factors.

$$b^2(y - k)^2 - a^2(x - h)^2 = a^2b^2$$

To convert from general form to standard form, use the process of completing the square.

Example 31:

Convert the following hyperbola from general form to standard form. Find the center point. Is the transverse axis horizontal or vertical?

$$4y^2 - x^2 + 24y + 4x + 28 = 0$$

Solution:

By now you're an old pro at completing the square, so we're not going to teach you again. Hyperbolas always involve subtraction. The y^2 is positive while the x^2 is negative, so we'll keep the y^2 first. This way the equation is clearly a subtraction problem.

$4y^2 - x^2 + 24y + 4x + 28 = 0$	We'll begin by grouping the terms that have a y together and the terms that have an x together.
$4y^2 + 24y - x^2 + 4x = -28$	Now we'll factor the 4 out of the y terms and the -1 out of the terms that have an x. Add the square of half the coefficients of first power terms to both sides.
$4(y^2 + 6y) - (x^2 - 4x) = -28$	
$4(y^2 + 6y + 9 - 9) - (x^2 - 4x + 4 - 4) = -28$	Move the extra constants to the other side. $4(-9) = -36$, $-1(-4) = 4$.
$4(y^2 + 6y + 9) - (x^2 - 4x + 4) = -28 + 36 - 4$	Factor.
$4(y + 3)^2 - (x - 2)^2 = 4$	Divide all terms by 4 to set the equation equal to 1.
$\dfrac{4(y + 3)^2}{4} - \dfrac{(x - 2)^2}{4} = \dfrac{4}{4}$	Reduce the common factors.
$\dfrac{(y + 3)^2}{1} - \dfrac{(x - 2)^2}{4} = 1$	Now that the hyperbola is in standard form, it's easy to see that the center point is $(2,-3)$. The transverse axis is vertical.

Example 32:

Find the center, the vertices, the foci, and the asymptotes of the following hyperbola.

$36x^2 - y^2 - 24x + 6y - 41 = 0$

Solution:

We'll begin by grouping the terms with an x together and the terms with a y together.

$36x^2 - y^2 - 24x + 6y - 41 = 0$

$36x^2 - 24x - y^2 + 6y = 41$

Factor 36 out of the x terms and -1 out of the y terms.

$36\left(x^2 - \dfrac{2}{3}x\right) - (y^2 - 6y) = 41$

$\left[\dfrac{1}{2}\left(-\dfrac{2}{3}\right)\right]^2 = \dfrac{1}{9}, \left[\dfrac{1}{2}(-6)\right]^2 = 9$

$36\left(x^2 - \dfrac{2}{3}x + \dfrac{1}{9} - \dfrac{1}{9}\right) - (y^2 - 6y + 9 - 9) = 41$

Distribute and move the extra constants to the other side of the equation.

$36\left(x^2 - \dfrac{2}{3}x + \dfrac{1}{9}\right) - (y^2 - 6y + 9) = 41 + 4 - 9$

Factor.

$36\left(x - \dfrac{1}{3}\right)^2 - (y - 3)^2 = 36$

Divide all terms by 36 to set the equation equal to 1.

$\dfrac{36\left(x - \dfrac{1}{3}\right)^2}{36} - \dfrac{(y - 3)^2}{36} = \dfrac{36}{36}$

Reduce the common factors.

$\dfrac{\left(x - \dfrac{1}{3}\right)^2}{1} - \dfrac{(y - 3)^2}{36} = 1$

Now that the equation is in standard form, the rest is easy. The transverse axis is horizontal.

The center point is $\left(\dfrac{1}{3}, 3\right)$, $a = 1$, $b = 6$. $c = \sqrt{a^2 + b^2} = \sqrt{1 + 36} = \sqrt{37} \approx 6.08$

The vertices are $\left(\dfrac{1}{3} \pm 1, 3\right)$, which are $\left(-\dfrac{2}{3}, 3\right)$ and $\left(\dfrac{4}{3}, 3\right)$. The foci are $\left(\dfrac{1}{3} \pm \sqrt{37}, 3\right)$, which are $\left(\dfrac{1}{3} - \sqrt{37}, 3\right)$ and $\left(\dfrac{1}{3} + \sqrt{37}, 3\right)$. The asymptotes are $y - 3 = \pm\dfrac{6}{1}\left(x - \dfrac{1}{3}\right)$, which are $y = -6x + 5$ and $y = 6x + 1$.

Before we ask you to try a few problems by completing the square, we would like to mention one last type of hyperbola. The last type of equation that represents a hyperbola is of the form $xy = c$, where c is a constant. The asymptotes are the coordinate axes, $x = 0$ and $y = 0$. The center point is the origin $(0,0)$.

Example 33:
Graph $xy = 4$.

Solution:

The easiest way to graph this hyperbola is to rewrite it by dividing both sides of the equation by x.

$xy = 4$

$\dfrac{xy}{x} = \dfrac{4}{x}$ Reduce the common factor of x on the left side of the equation.

$y = \dfrac{4}{x}$

If you have had precalculus, you'll recognize this as the reciprocal function. The only value x can't be is 0, because division by 0 is undefined. Therefore, this function has the vertical asymptote $x = 0$. The degree of the numerator, 0, is less than the degree of the denominator, 1, so the horizontal asymptote is $y = 0$. When the values for x are positive, the corresponding values for y will also be positive, because 4 divided by a positive number is positive. When the values for x are negative, the values for y will be negative, because 4 divided by a negative number is negative. As the positive x values increase, the value of y approaches 0. As the positive x values approach 0, the y values increase without bound. As the negative x values approach 0, the y values decrease without bound. As the negative x values decrease, their corresponding y values approach 0. The following table of values illustrates this. These problems are covered in more detail in our precalculus book, also published by Wiley.

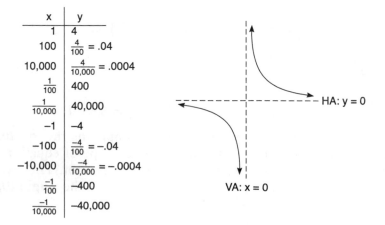

x	y
1	4
100	$\frac{4}{100} = .04$
10,000	$\frac{4}{10,000} = .0004$
$\frac{1}{100}$	400
$\frac{1}{10,000}$	40,000
−1	−4
−100	$\frac{-4}{100} = -.04$
−10,000	$\frac{-4}{10,000} = -.0004$
$\frac{-1}{100}$	−400
$\frac{-1}{10,000}$	−40,000

Now it's your turn to try a few problems.

1. Find the center point of the following hyperbola.
$4x^2 - y^2 + 8x - 4y - 4 = 0$
2. Find the center, the foci, and the vertices of the following hyperbola.
$9x^2 - 4y^2 + 54x + 8y + 41 = 0$
3. Find the center point, vertices, foci, and asymptotes and sketch the graph of the following hyperbola.
$9y^2 - 4x^2 - 18y + 24x - 63 = 0$
4. Find the center point, vertices, foci, and asymptotes and sketch the graph of the following hyperbola.
$x^2 - y^2 - 2x - 4y = 4$
5. Find the center point of the following hyperbola. Is the transverse axis vertical or horizontal?
$y^2 - x^2 - 6x - 8y - 29 = 0$
6. Graph the following hyperbola.
$xy = 16$
7. Graph the following hyperbola.
$xy = -4$

ANSWERS

1. $4x^2 - y^2 + 8x - 4y - 4 = 0$
$4x^2 + 8x - y^2 - 4y = 4$
$4(x^2 + 2x) - (y^2 + 4y) = 4$
$4(x^2 + 2x + 1) - (y^2 + 4y + 4) = 4 + 4 - 4$
$4(x + 1)^2 - (y + 2)^2 = 4$
$\dfrac{(x + 1)^2}{1} - \dfrac{(y + 2)^2}{4} = 1$ The center point is $(-1,-2)$

2. $9x^2 - 4y^2 + 54x + 8y + 41 = 0$
$9(x^2 + 6x) - 4(y^2 - 2y) = -41$
$9(x^2 + 6x + 9) - 4(y^2 - 2y + 1) = -41 + 81 - 4$
$9(x + 3)^2 - 4(y - 1)^2 = 36$
$\dfrac{(x + 3)^2}{4} - \dfrac{(y - 1)^2}{9} = 1$

The center point is $(-3,1)$. The transverse axis is horizontal. $a = 2$, $b = 3$. The vertices are $(-1,1)$ and $(-5,1)$.

$c = \sqrt{4 + 9} = \sqrt{13} \approx 3.61$

The foci are $(-3 - \sqrt{13},1)$ and $(-3 + \sqrt{13},1)$.

The asymptotes are $y - 1 = \pm\dfrac{3}{2}(x + 3)$, which are $y = \dfrac{3}{2}x + \dfrac{11}{2}$ and $y = -\dfrac{3}{2}x - \dfrac{7}{2}$.

3. $9y^2 - 4x^2 - 18y + 24x - 63 = 0$

$9(y^2 - 2y) - 4(x^2 - 6x) = 63$

$9(y^2 - 2y + 1) - 4(x^2 - 6x + 9) = 63 + 9 - 36$

$9(y - 1)^2 - 4(x - 3)^2 = 36$

$$\frac{9(y - 1)^2}{36} - \frac{4(x - 3)^2}{36} = \frac{36}{36}$$

$$\frac{(y - 1)^2}{4} - \frac{(x - 3)^2}{9} = 1$$

The transverse axis is vertical. The center point is $(3,1)$. $a = 2$, $b = 3$, $c = \sqrt{4 + 9} = \sqrt{13} \approx 3.61$. The vertices are $(3,3)$ and $(3,-1)$. The foci are $(3, 1 - \sqrt{13})$ and $(3, 1 + \sqrt{13})$. The asymptotes are $y = \frac{2}{3}x - 1$, $y = -\frac{2}{3}x + 3$.

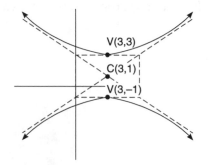

4. $x^2 - y^2 - 2x - 4y = 4$

$(x - 1)^2 - (y + 2)^2 = 4 + 1 - 4$

$(x - 1)^2 - (y + 2)^2 = 1$

C: $(1,-2)$, $a = 1$, $b = 1$, $c = \sqrt{2} \approx 1.41$

V: $(0,-2)$ and $(2,-2)$, F: $(1 - \sqrt{2},-2)$ and $(1 + \sqrt{2},-2)$.

The transverse axis is horizontal.

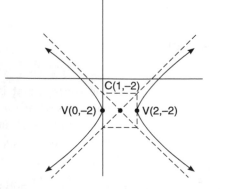

5. $y^2 - x^2 - 6x - 8y - 29 = 0$

$(y^2 - 8y + 16 - 16) - (x^2 + 6x + 9 - 9) = 29$

$(y - 4)^2 - (x + 3)^2 = 29 + 16 - 9$

$(y - 4)^2 - (x + 3)^2 = 36$

$\dfrac{(y - 4)^2}{36} - \dfrac{(x + 3)^2}{36} = 1$ C: (–3,4). The transverse axis is vertical.

6. $xy = 16$

$y = \dfrac{16}{x}$

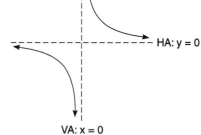

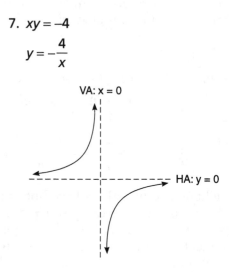

7. $xy = -4$

$y = -\dfrac{4}{x}$

Applications

In this section we'll apply our knowledge of conic sections to real–life applications. First it's important to make sure that you can tell the difference between the equations of the conic sections. See if you can tell the shape of the following equations: circle, ellipse, hyperbola, or parabola.

1. $x^2 + 4y^2 + 6x - 8y + 9 = 0$

2. $x^2 = -6y$

3. $y^2 + 2x + 8y + 13 = 0$

4. $4x^2 - y^2 = 4$

5. $x^2 + y^2 = 100$

6. $9x^2 - y^2 - 9 = 0$

7. $3x^2 + 4y^2 = 12$

8. $\dfrac{4x^2}{9} + \dfrac{y^2}{16} = 1$

9. $4x^2 - 12x + 12y + 7 = 0$

10. $x^2 + y^2 - 6x + 4y - 3 = 0$

Before we give you our answers to check with yours, here's a list of the standard forms of the conic sections we have covered in this chapter.

circle: $(x - h)^2 + (y - k)^2 = r^2$

parabola: $(y - k)^2 = 4p(x - h)$ $(x - h)^2 = 4p(y - k)$

ellipse: $\dfrac{(x - h)^2}{a^2} + \dfrac{(y - k)^2}{b^2} = 1$ $\dfrac{(x - h)^2}{b^2} + \dfrac{(y - k)^2}{a^2} = 1$

hyperbola: $\dfrac{(x - h)^2}{a^2} - \dfrac{(y - k)^2}{b^2} = 1$ $\dfrac{(y - k)^2}{a^2} - \dfrac{(x - h)^2}{b^2} = 1$

Notice the equations of a circle, ellipse, and hyperbola have both the x and the y squared. The equation of a parabola has the x or the y squared, not both. The equations of a circle and an ellipse have a + sign between the x squared and the y squared terms. The ellipse is set equal to 1 and can have denominators other than 1. The circle doesn't have to be set equal to 1 and doesn't show denominators. The equation of the hyperbola has the x squared and the y squared separated by a − sign. To figure out which type of conic section the given equations represent, you may have to complete the square.

1. $\dfrac{(x+3)^2}{4} + (y-1)^2 = 1$ ellipse

2. parabola

3. parabola

4. hyperbola

5. circle

6. hyperbola

7. ellipse

8. ellipse

9. parabola

10. $(x-3)^2 + (y+2)^2 = 16$ circle

Let's put all the knowledge you've acquired in this chapter to some practical use. In the previous sections we showed you how to write the equations of circles, ellipses, parabolas, and hyperbolas. We also showed you how to graph these figures. By now you should be able to tell the difference between the equations of these conic sections. Now it's time to apply your newfound knowledge to some word problems that apply to the real world. We're not saying all of these problems are of the type an average person would use on a daily basis, but they are fun and are typical problems that might be used by an engineer, an architect, a scientist, a ship's captain, a doctor, and others.

What we're going to do now is try a few applied problems so you can get a feel for the way in which conic sections can appear and be solved in real-life situations.

Example 34:

A semielliptical archway over a one-way road has a height of 10 feet and a width of 40 feet (see the following figure). Your truck has a height of 9 feet and a width of 10 feet. Will your truck clear the opening of the archway?

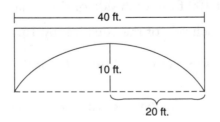

Solution:

To determine the clearance, we must find the height of the archway 5 feet from the center. If this height is 9 feet or less, the truck will not clear the opening. In the following figure we've constructed a coordinate system with the *x*-axis on the ground and the origin at the center of the archway. Also shown is the truck whose height is 9 feet.

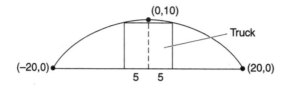

Using the standard form of the equation of an ellipse centered at the origin, $\dfrac{x^2}{a^2} + \dfrac{y^2}{b^2} = 1$, we can express the equation of the archway as $\dfrac{x^2}{20^2} + \dfrac{y^2}{10^2} = 1$ or $\dfrac{x^2}{400} + \dfrac{y^2}{100} = 1$.

The middle of the back of the truck is located at the origin; therefore, the width of the truck is 5 feet on either side of the origin, so $x = 5$. We have to find the height of the archway 5 feet from the center to determine whether the truck can clear the archway. To do this we'll substitute 5 for x and solve for y.

$\dfrac{x^2}{400} + \dfrac{y^2}{100} = 1$	Substitute 5 for x.
$\dfrac{5^2}{400} + \dfrac{y^2}{100} = 1$	
$\dfrac{25}{400} + \dfrac{y^2}{100} = 1$	Reduce the fraction $\dfrac{25}{400} = \dfrac{1}{16}$.
$\dfrac{1}{16} + \dfrac{y^2}{100} = 1$	Multiply all terms by 1,600 to clear out the fractions.
$1{,}600\left(\dfrac{1}{16}\right) + 1{,}600\left(\dfrac{y^2}{100}\right) = 1(1{,}600)$	
$100 + 16y^2 = 1{,}600$	Subtract 100 from each side of the equation.
$16y^2 = 1{,}500$	Divide each side of the equation by 16.
$y^2 = \dfrac{1{,}500}{16}$	Take the square root of both sides of the equation.
$y = \sqrt{\dfrac{1{,}500}{16}} \approx 9.68$	Now we can see that the truck will clear the archway.

Example 35:

The towers of the Golden Gate Bridge connecting San Francisco to Marin County are 1,280 meters apart and rise 160 meters above the road. The cable between the towers has the shape of a parabola, and the cable just touches the sides of the road midway between the towers. What is the height of the cable 200 meters from a tower?

Solution:

We'll begin by drawing the given information with the center of the cable located at the origin.

The 1,280 meters is equally divided on both sides of the origin, therefore we can assume $x = 640$. When $x = 640$, $y = 160$. This parabola, which is centered at the origin, opens up, so we'll use the formula, $x^2 = 4py$. We know $x = 640$ when $y = 160$. We'll use this knowledge to solve for p.

$x^2 = 4py$ Substitute 640 for x and 160 for y.

$(640)^2 = 4p(160)$ Divide both sides of the equation by $4(160)$ to solve for p.

$\dfrac{(640)^2}{4(160)} = p$ Reduce.

$p = 640$

Now that we know the value of p, we'll substitute it into $x^2 = 4py$.

$x^2 = 4(640)y = 2{,}560y$

To find the height of a cable 200 meters from a tower, we substitute 440 (640 − 200) into the formula and solve for x.

$x^2 = 2{,}560y$ Substitute 440 for x.

$(440)^2 = 2{,}560y$ Divide both sides of the equation by 2,560.

$\dfrac{(440)^2}{2{,}560} = y$ Simplify.

$y \approx 75.625$ meters

Example 36:

A draftsman is drawing a friction drive in which two circular disks are in contact with each other. They are represented by circles in the following drawing. The first has a radius of 10.0 cm and the second has a radius of 12.0 cm. What is the equation of each circle if the origin is the center of the first circle and the positive x-axis passes through the center of the second circle?

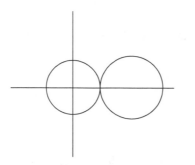

Solution:

Because the center of the smaller circle is at the origin, we can use the following equation to write the equation of the smaller circle.

$$x^2 + y^2 = r^2$$

$$x^2 + y^2 = 10^2$$

$$x^2 + y^2 = 100$$

The fact that the two disks are in contact tells us that they meet at the point (10,0). Knowing that the radius of the larger circle is 12.0 cm tells us that the center is at (22,0). We'll use the following equation to write the equation of the larger circle.

$$(x - h)^2 + (y - k)^2 = r^2$$

$$(x - 22)^2 + (y - 0)^2 = 12^2$$

$$(x - 22)^2 + y^2 = 144$$

Example 37:

A parabolic reflector is to be designed with a light source at its focus, $2\frac{1}{4}$ inches from its vertex. If the reflector is to be 10 inches deep, how far will the outer rim be from the source? See the following figure.

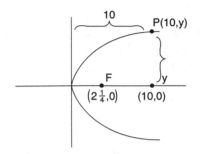

Solution:

Fortunately we've supplied you with a diagram of the word problem. Notice we have used the origin as the vertex of the parabola. Sometimes you will have to draw your own diagram. When this happens we suggest you use the origin as the vertex whenever possible. Because the parabola is horizontal, the *y* is squared, not the *x*. The standard equation is $y^2 = 4px$.

Because $p = 2\frac{1}{4}$, $y^2 = 4\left(2\frac{1}{4}\right)x = 4\left(\frac{9}{4}\right)x = 9x$. The equation of the parabolic

cross-section is $y^2 = 9x$. As the reflector is to be 10 inches deep, we can designate a point on its outer rim as $(10,y)$. Substituting these coordinate values in the equation $y^2 = 9x$, we get $y^2 = 9(10) = 90$, or $y = 9.486$ inches, and the total breadth is $2y = 2(9.486) = 18.972$ inches. To find the focal radius (that is, the distance of the point *P* from the focus *F*), we can use the relationship $FP = x + p = 10 + 2\frac{1}{4} = 12.25$ inches, because, by the definition of a parabola, the focal radius to any point on the curve is equal to the distance of the same point from the directrix.

Example 38:

The Moon travels about Earth in an elliptical orbit with Earth at one focus, as shown in the following illustration. The major and minor axes of the orbit have lengths of 768,806 kilometers and 767,746 kilometers, respectively. Find the greatest and smallest distances from Earth's center to the Moon's center.

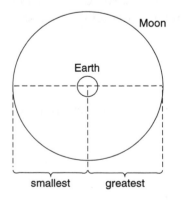

Solution:

The length of the major axis is $2a$, and the length of the minor axis is $2b$. $2a = 768,806$ and $2b = 767,746$; $a = 384,403$ and $b = 383,873$. To find the length of c, we'll use the formula $c = \sqrt{a^2 - b^2} = \sqrt{384,403^2 - 383,873^2} \approx 20,179$. Therefore, the greatest distance between the center of Earth and the center of the Moon is $a + c \approx 404,582$ kilometers, and the smallest distance is $a - c \approx 364,224$ kilometers.

Example 39:

Two microphones, 1 mile apart, record an explosion. Microphone A receives the sound 2 seconds before microphone B. Where did the explosion occur?

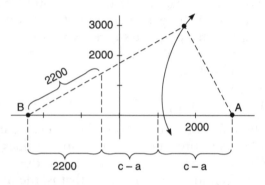

Solution:

Assuming sound travels at 1,100 feet per second, you know that the explosion took place 2,200 feet farther from B than from A, as shown in the previous figure. The locus of all points that are 2,200 feet closer to A than to B is one branch of the hyperbola $\dfrac{x^2}{a^2} - \dfrac{y^2}{b^2} = 1$, where $c = \dfrac{5,280}{2} = 2,640$ and $a = \dfrac{2,200}{2} = 1,100$. Thus, $b^2 = c^2 - a^2 = 5,759,600$, and you conclude that the explosion occurred somewhere on the right branch of the hyperbola given by $\dfrac{x^2}{1,210,000} - \dfrac{y^2}{5,759,600} = 1$.

Now that you've seen a few word problems involving conic sections, it's your turn to try a few problems.

SELF-TEST 9

For problems 1 through 4, identify which type of conic section the given equations are.

1. $\dfrac{(x-3)^2}{4} - \dfrac{(y+2)^2}{9} = 1$

2. $(y-2)^2 = 8(x-1)$

3. $\dfrac{(x+4)^2}{9} + \dfrac{(y+1)^2}{4} = 1$

4. $x^2 + y^2 + 4x - 6y + 12 = 0$

5. A simply supported beam is 64 feet long and has a load at the center. The deflection of the beam at its center is 1 inch. The shape of the deflected beam is parabolic.

   a. Write an equation of the parabola.

   b. How far from the center of the beam is the deflection $\dfrac{1}{2}$ inch?

6. An arch in the shape of the upper half of an ellipse is used to support a bridge that is to span a river 20 meters wide. The center of the arch is 6 meters above the center of the river. Write an equation for the ellipse in which the x-axis coincides with the water level and the y-axis passes through the center of the arch. Find the height of the archway 5 feet from the center.

7. The arch of a bridge over a highway is semielliptic. The base of the bridge covers the entire width of the highway, and the highest part of the bridge is 20 feet directly above the center of the highway. What is the height of the bridge 10 feet from the center of the road?

8. A satellite to study Earth's atmosphere has a minimum altitude of 600 miles and a maximum altitude of 2,000 miles. If the path of the satellite around Earth is an ellipse with the center of Earth at one focus, what is the equation of its path? Assume the radius of Earth is 4,000 miles.

9. A satellite dish is shaped like a paraboloid of revolution. The signals that emanate from a satellite strike the surface of the dish and are reflected to a single point, where the receiver is located. If the dish is 8 feet across at its opening and is 3 feet deep at its center, at what position should the receiver be placed?

ANSWERS

1. hyperbola

2. parabola

3. ellipse

4. circle $(x + 2)^2 + (y - 3)^2 = 1$

5. Let's draw this information with the center of the beam at the origin.

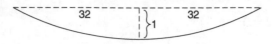

We'll use the standard equation of a parabola that opens up with a vertex located at the origin.

a. $x^2 = 4py$ Substitute 32 for x and 1 for y.

    $32^2 = 4p(1)$ Solve for p.

    $1,024 = 4p$ Divide both sides of the equation to solve for p.

    $256 = p$ Substitute 256 for p into the standard equation.

    $x^2 = 4(256)y$

    $x^2 = 1,024y$

b. Substitute $\dfrac{1}{2}$ for y and solve for x.

    $x^2 = 4(256)\left(\dfrac{1}{2}\right)$

    $x^2 = 512$ Take the square root of both sides of the equation.

    $x = \sqrt{512} \approx 22.627$ feet from the center.

6.

$\dfrac{x^2}{a^2} + \dfrac{y^2}{b^2} = 1$ Substitute 10 for a and 6 for b.

$\dfrac{x^2}{100} + \dfrac{y^2}{36} = 1$ Substitute 5 for x then solve for y.

$\dfrac{25}{100} + \dfrac{y^2}{36} = 1$ Multiply all terms by the LCD 3,600.

$3,600\left(\dfrac{25}{100}\right) + 3,600\left(\dfrac{y^2}{36}\right) = 3,600(1)$ Reduce the common factors.

$36(25) + 100y^2 = 3,600$

$100y^2 = 2,700$ Divide both sides by 100.

$y^2 = 27$ Take the square root of both sides of the equation.

$y = \sqrt{27} \approx 5.196$ feet

7. First we'll draw a diagram to represent the given information.

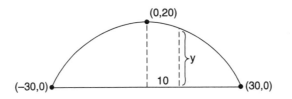

We'll begin by using the standard form of an ellipse centered at the origin. Substitute in the values for *a* and *b,* then solve for *y.*

$$\frac{x^2}{a^2} + \frac{y^2}{b^2} = 1$$ Substitute 30 for *a* and 20 for *b.*

$$\frac{x^2}{30^2} + \frac{y^2}{20^2} = 1$$

$$\frac{x^2}{900} + \frac{y^2}{400} = 1$$ Substitute 10 for *x* and solve for *y.*

$$\frac{100}{900} + \frac{y^2}{400} = 1$$ $\frac{100}{900} = \frac{1}{9}$ Subtract $\frac{1}{9}$ from both sides of the equation.

$$\frac{y^2}{400} = 1 - \frac{1}{9}$$ $1 - \frac{1}{9} = \frac{9}{9} - \frac{1}{9} = \frac{8}{9}$

$$\frac{y^2}{400} = \frac{8}{9}$$ Multiply both sides of the equation by 400.

$$y^2 = 400\left(\frac{8}{9}\right)$$ Take the square root of both sides of the equation.

$$y = \sqrt{\frac{3{,}200}{9}} \approx \frac{40\sqrt{2}}{3} \approx 18.862 \text{ feet}$$

8. First we'll draw a diagram to represent the given information.

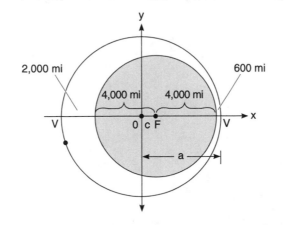

$2a = 2{,}000 + 4{,}000 + 4{,}000 + 600 = 10{,}600$ miles

$a = 5,300$ miles

From the right focus to the right vertex is 4,600 miles

$c = a - 4,600 = 5,300 - 4,600 = 700$ miles

$b^2 = a^2 - c^2 = 5,300^2 - 700^2 = 2.76 \times 10^7$ mi^2

$a^2 = 5,300^2 = 2.81 \times 10^7$

The equation is $\dfrac{x^2}{2.81 \times 10^7} + \dfrac{y^2}{2.76 \times 10^7} = 1.$

9. As usual, we'll begin by drawing the conic section centered at the origin. A paraboloid is a three-dimensional parabola.

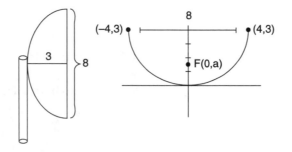

| $x^2 = 4py$ | Substitute 4 for x and 3 for y. |
| $4^2 = 4p(3)$ | Solve for p. |

$16 = 12p$

$p = \dfrac{4}{3} = 1\dfrac{1}{3}$ feet

The receiver should be $1\dfrac{1}{3}$ feet from the base of the dish along its axis of symmetry.

Index